PRINCIPLES OF SECURITY

SECOND EDITION

Truett A. Ricks, CPP
Dean,
College of Law Enforcement,
Eastern Kentucky University

B.G. Tillett, CPP
Professor and Chair,
Department of Loss Prevention and Safety,
Eastern Kentucky University

Clifford W. VanMeter, Ph.D.
Director,
Police Training Institute,
University of Illinois

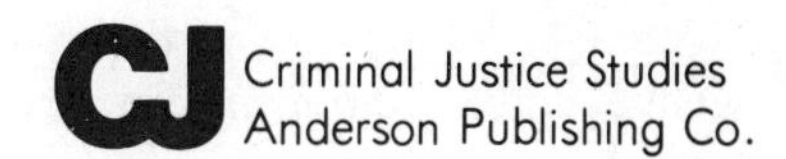
Criminal Justice Studies
Anderson Publishing Co.

PRINCIPLES OF SECURITY Second Edition

Second Printing–February 1988

Library of Congress Cataloging-in-Publication Data
Ricks, Truett A.
Principles of security.

(Criminal justice studies)
Includes bibliographical references and index.
1. Private security services–United States.
I. Tillett, B.G. (Bill G.) II. Van Meter, Clifford W. III. Title.

HV8290.R5 1987 362.8'8 87-12159

ISBN 0-87084-745-7

Kelly Humble *Managing Editor*

Table of Contents

Preface Introduction

Six years ago when we wrote the first edition, we prepared a preface in which we stated our objectives and focus for the book. In reviewing the preface, we find that the observations we made are as relevant today as we felt they were six years ago. Thus, other than the comments contained in this paragraph, the preface to our second edition is the same as the first edition.

June, 1987

Preface

In recent years it has become increasingly apparent that public law enforcement agencies cannot provide the needed resources nor the manpower to protect private property. Thus, the role of private security has become more visible and important as the reality of self-protection and loss prevention has been recognized and accepted by the businessman and property owner.

It is the objective of this book to provide the reader with the basic principles of security and loss prevention that are common and fundamental to all areas of business and asset protection. After presenting a brief look at the history of security, the text provides coverage on the security industry, the threat environment, risk analysis, fundamentals of physical security, common security problems, safety and accident prevention, and the basic elements of fire prevention and protection.

The materials and information that are presented in this text will serve to provide an introduction to a field of study that is undergoing rapid change and enlargement. It is designed for those students interested in a career in security, for those in other fields of study incorporating loss prevention activities such as manufacturing, commerce, finance, health care, national defense, government, architecture, law enforcement, and so on, and for the security practitioner who needs to check, evaluate, or apply the fundamental principles of this text to a particular situation.

In general, private security will progress according to the quality of information and personnel that are available and utilized. If properly educated, trained, and selected, professional security personnel will be the rule rather than the exception. We hope this text will serve to provide a needed step in that direction.

Truett A. Ricks
Bill G. Tillett
Clifford W. Van Meter

January, 1981

Acknowledgments

Writing of this text has been a rewarding experience for the authors from many standpoints. First, it has provided us an opportunity to continue our research in the ever-changing and dynamic field of private security. As the research was conducted we had the challenge of bringing current and reliable information to our readers. Second, it forced us to work closely with our academic colleagues at Eastern Kentucky University and the University of Illinois in reviewing our manuscripts from a critical yet supportive point of view.

However, as in any undertaking of this scope, there are several people who provided extra assistance and support. Mr. Clarence H. A. Romig, CPP, Adjunct Faculty Member, University of Illinois Police Training Institute provided that extra assistance by writing Chapters 10 and 14. Dr. Larry K. Gaines, Associate Professor, Ann Duerson, Joyce Simpson, Verna Casey, and Priscilla Ray of the College of Law Enforcement at Eastern Kentucky University provided invaluable editorial, research, and clerical assistance. And the authors would like to acknowledge the support and patience provided by their families: Judy, Bill, and Tracey Van Meter; Linda, Terri, Tim, and Lori Tillett; Betty and Tammy Ricks.

Chapter 1
Historical Development of Security

Introduction

The history of man can be seen as a series of mankind's attempts to provide for his safety, security, and freedom from fear and danger. Mankind has developed weapons, built barriers around dwellings, and devised codes of conduct to protect property and welfare.

Many of these efforts have been uncovered by archaeologists, and some are still visible and even functional after hundreds of years. They are illustrative of man's attempt to isolate, separate or secure himself from others. Evidence of ancient *pole* dwellings has been uncovered where houses and even entire villages were constructed on lakes. Even today this method of home building is still apparent in certain parts of Africa and the Far East. Cave dwellings, often high on cliffs, indicate that early man utilized whatever was natural to his environment for his protection. Many ancient cities have been uncovered to reveal high walls that surrounded the entire town. The Great Wall of China was built centuries ago by the Chinese to keep out the Mongols. Throughout Europe are hundreds of medieval castles encircled by moats, and connnected to land by drawbridges. In the United States, the historic movement westward required the construction of forts to provide security for the early settlers from hostile Indians.

History illustrates the ability of mankind to innovate and adjust to changes in the environment. Security needs were first recognized and accomplished by the individual and his family, but ultimately his basic needs became the same as those of the community and the nation. Social customs and relationships expanded beyond the ties of kinship until the informal activities of early man became the formal regimentations and rules of a larger, more complex society.

Early European Developments

The beginnings of formal security activites developed from the crude and unsophisticated survival tactics utilized by early family groups and the more elaborate tactics utilized once families began to form tribes or clans. The small family groups blended into larger groups where organizational unity was reinforced by kinship and common traditions, customs, language, and common interests. The entire tribe or clan assumed a collective responsibility for taking care of the family, the tribe, and the village. Tribal security customs developed from this rudimentary system which provided security for the individual and the group.

Historically, security responsibilities have followed social and economic developments and the basic principles of law and justice. The culture of the prehistoric hunter and food gatherer limited security to the safety and integrity of persons, their social arrangements within the group and their few possessions, which consisted of limited provisions and a few tools of stone and bone. With the advent of agriculture and the domestication of animals, security took on new dimensions. The cultivation of plants added the need for continuous possession of land. Instead of roaming the land in search of food and shelter, man settled down to build shelters for himself and his animals. The land, animals, buildings, and crops became coveted possessions with economic value. Throughout the ancient world, the dealings in agriculture, commerce, the crafts, and the professions became regulated as societal rules and codes evolved to protect life and property in a changing agricultural society with an increasingly complex social order.

The Anglo-Saxons

Around 400 A.D., as land began to be scarce on the European continent, the Anglo-Saxons, a people of German origin, began to migrate to England in large numbers. They brought a security system of compulsory communal responsibility for protecting and taking care of the security needs of individuals, families, tribes and villages. Under this system the people were organized into groups of ten families, or householders called a *tithing*, and further into groups of ten tithings called a *hundred*. Each tithing selected a *tithingman* to represent the group. Ten tithingmen represented a hundred and had a King's *reeve* to speak for them. Several hundreds made up a *shire* (a geographical administrative district) and a *shire-reeve* (the title *sheriff* is derived from this office) was the spokesman for the entire shire. Several shires constituted an *earldom*, headed by an *earldom-man*.

The feudal tithing system stressed social stability and hierarchical control. It was considered the duty of every citizen to be a policeman. The members of the group were themselves responsible for whatever offenses were committed within their borders. The tithingman who was elected from the group was given the responsibility for calling the group to action, i.e., raising the "hue and cry," and meting out punishment. The English common law's process, wherein every able-bodied man was required to join in the communal pursuit of law breakers, was the origin of citizen's arrest.

The basic economic and societal ties of feudalistic society were kinship and the relationship between an individual and his landlord. The kinship philosophy required the kindred to seek vengeance and compensation for an act against a relative. The servant-landlord philosophy was a bilateral (though unequal) agreement: the landlord would protect his servant from outside forces, and in return the servant would till the land to provide the necessary substances for life.

The Norman Era

The dual system of the landlord-servant relationship and strength of kindred justice was completely changed in 1066, when William, Duke of Normandy, invaded and conquered England. William dispossessed the old English nobility and initiated a comprehensive political, economic, and social survey of England. A national system which placed emphasis on collective and community security at the expense of individual freedom was initiated. William placed England under martial law, divided the country into fifty-five military districts and placed a *tenant-in-chief* in charge of each district. The Anglo-Saxon system of security through shire-reeves and courts of shires was modified as judicial processes were centralized under the King's judges who traveled throughout the country. These traveling judges were the forerunners of modern day circuit judges.

Around 1100 A.D. the office of *constabuli* was established within the shire system. The word was derived from the Latin *comes stabuli* meaning an officer of the stable. A constable was appointed to every hundred to aid the reeve in the conduct of his duties.

In 1116 A.D., Henry I, son of William the Conquerer, issued the *Leges Henrici* in which he gave himself the title of Law Giver. The *Leges Henrici* originated the idea of a separation between those crimes judged to be serious or felonious and the lesser offenses deemed to be misdemeanors. If a felony was committed, both officers of the crown and citizens had equal authority to make an arrest for crimes committed in their presence. In the case of misdemeanors only officers of the crown had the right to arrest.

The Assize of Clarendon in 1166 revived the Anglo-Saxon system of mutual security or *frankpledge*. A section of this code established the *grand jury*, and initiated the end of the trial-by-ordeal and trial-by-combat. The creation of the jury system brought about a change in the fundamental concepts of justice by establishing rules of evidence and new protections for the rights of individuals and the security of persons and property.

English Reforms and Change (1200-1700)

King John, a hated ruler of England, was forced by the barons to sign the Magna Carta (Great Charter) in 1215. This document established a clear separation between local and national government and established the principle that the King was subject to the law. The Magna Carta also promulgated *due process*, the course of legal proceedings to protect individual rights and liberties. It served to strengthen the importance and role of local grand juries, circuit judges, coroners, and justices of the peace.

William the Conqueror's innovative national security system had deteriorated by the time the Statute of Winchester (also known as the Statute of Westminster), issued in 1285, reestablished a formalized law enforcement system throughout England. This document required that every area of England implement a security force, which was specified according to time, place and number of personnel. It established a system of patrolling called *watch and ward*. Every district was to control crimes within its boundaries; the gates of all towns were required to be closed at dusk, and all persons not residing in the town were required to check in with local authorities. The night watch and the office of bailiff were initiated by the Statute. Bailiffs checked on strangers and lodgers at inns in the town, and guards manned the city gates from sunset to sunrise to secure the city. Additionally, these *watchmen* grouped into a marching watch to limit the movement of townspeople during certain hours, establishing the concepts of mobile patrol and curfew as security measures. Every able male of the community had to serve his turn on the watch and, like the constables, members of the watch were unpaid.

A slow but continuous change in pattern of the European economy from the fourteenth to the end of the seventeenth century, termed the commercial revolution, contained many changes and trends including increased trade, exploration, and the rise of the merchant class. Advances in transportation expanded trade throughout the known world. An elaborate system of international markets, trade, and colonization required increased productive capacity and surplus from agriculture, animal husbandry, and the skilled trades.

Large land holdings were essential to the production of economically feasible amounts of agricultural products. The landlords consolidated the small holdings of the peasants, who were excluded by enclosure acts from open grazing privileges, into large *manor* farms. Tenants displaced by the consolidation of lands migrated to the cities. Cultural patterns and family traditions were in upheaval as the constraints of medieval society were disintegrating. Mass unemployment, poverty, health and welfare needs caused social unrest and dynamic changes and increases in security problems. There was no civil force which could effectively protect persons and property. The protection of production goods while in storage and transit became a critical concern of the merchant class. To combat the problem, individual merchants and tradesmen hired men to guard their buildings, shops, stores of goods, and caravans. The use of the forerunners of private detectives to locate and identify stolen property began during this period. The parochial police also came into existence, as many English cities arranged in distinct population and geographic districts by religion or ethnic background hired their own police to protect them and their property. These private police performed essentially the same function as would a police officer except that their responsibility extended only to the boundary of the district.

Thus, for some 300 years the cities and countryside of England were policed by a fragmented system of constables and watchmen. The first police officials were the shire-reeve (sheriff) and, later, the parish constable. The parish constables were charged with maintaining law and order, and were responsible to the Justice of the Peace. Since there was no central authority coordinating this justice system, justice was often fragmented and of inferior quality, and there was a great deal of corruption within the system. The Justices of the Peace, appointed by the Crown, were unpaid and usually selected from the gentry. The appointees frequently used their positions to better their own needs, maintaining the status quo by enforcing the laws in favor of their own social class.

Prior to 1737, all personnel who received pay for watch duty were paid exclusively by individuals or private groups. In that year George II began to pay watchmen with tax monies collected specifically for security protection.

Thus, the period of the commercial revolution was one of great turmoil and social upheaval, and provided the first real evidence that the English public protective system was unable to cope with the task of providing a minimal amount of protection for emerging business and commercial enterprises.

The Eighteenth Century

Eighteenth Century England, leading the way toward the Industrial Revolution, saw an almost complete breakdown of the constable system of law enforcement. The rural population began to move to the cities to find jobs. Weaving and knitting machines and new methods of metal production resulted in enormous industrial progress and the creation of previously nonexistent jobs. The capacity to produce expanded as never before, yet there was also poverty and suffering among masses of people. Crime grew to alarming proportions as the displaced, the poor, and disoriented increased in number.

The noted English author, Patrick Pringle, in *Hue and Cry*, had the following to say about Eighteenth Century England:

> No one thought our policemen wonderful in the eighteenth century. What struck foreigners as remarkable was that we had none. London was the greatest town in the world; it was also the most lawless. This was not because the British were too soft with criminals, although being British they naturally thought they were. In fact, Britain not only had the most criminals; it also had the harshest Criminal Code. Men, women, and children were liable to be hanged for offences that in other countries were considered quite trivial: associating with gypsies or cutting hop-binds, for example, or entering land with intent to kill rabbits; impersonating a Chelsea Pensioner, or chipping bits out of Westminster Bridge. A boy or girl of seven could be sentenced to death for stealing a pocket handkerchief. The law got steadily harsher throughout the century, while in other countries it was getting steadily more humane . . .
>
> When England emerged from feudalism she did not need a standing army, for she had no land frontiers; and one result of this geological accident was that England had no men-of-arms to use as police. Her only security legacy from the feudal era was the amateur parish-constable system. This continued to work adequately in country parishes, but it was not suited to larger units, such as towns. By 1700 the population of the metropolis (London), as it was called, was well over half a million. During the first half of the century it hardly increased at all, in spite of considerable immigration from the country; for deaths greatly exceeded births. Between 1740 and 1742, for example, there were twice as many burials as baptisms. The main check on the growth of population was the wholesale murder of children by their parents and parish authorities. After a careful investigation, Jonas Hanway estimated that over 75 per cent of all children died before they were five, and that infant mortality among illegitimate children was over 95 per cent. Some illegitimate children were murdered at birth by their mothers or, more commonly, laid out in the streets to die of exposure; others starved to death in workhouses or in the care of nurses who specialized in taking them off the hands of the parish. Some nurses, however, kept children alive to use them for begging after first blinding or maiming them to increase their value.[1]

Such was the world of Jonathan Wild. In 1743, Henry Fielding, later to become the chief magistrate of the Bow Street area of London, wrote a political satire entitled, *Life of Mr. Jonathan Wild, The Great.* While not a factual biography of the life and times of Jonathan Wild, the events of Fielding's book were drawn from Wild's career as one of England's most notable criminals. Fielding, through the personage of Wild, characterized the aggregate nature of crime and justice in Eighteenth Century England. The corruption and ineptness of the English system of justice was evidenced in Wild's criminal ventures as a fence, smuggler, thief taker and criminal mastermind.

In 1748, when Henry Fielding became chief magistrate of Bow Street, crime had become rampant, counterfeit money was more common than good money, and there were over 100 offenses punishable by death. Fielding set himself two tasks: to eliminate existing crime, and to prevent fresh outbreaks in the future. To achieve these aims he considered three things necessary: the active cooperation of the public, a stronger police, and the removal of the causes of crime and of conditions in which it flourished. In his five years at Bow Street, Fielding's significant contributions included a foot patrol to make the streets safe, a mounted patrol for the highways, the Bow Street Runners (special investigators), and police courts.

Fielding's scheme was to thwart criminals by actively seeking them out and investigating their activities:

> Citizens, he (Fielding) realized, might combine together collectively, to go into the streets, trace the perpetrators of crimes in their haunts and meet the instigators of mob gatherings before they had assembled a following and caused destruction. He saw that it was possible to prevent, instead of repressing crime and disorder . . .[2]

This was in complete contrast to the constables and watchmen who could not be found when trouble erupted.

Fielding formed a band of volunteers who arrested numerous criminals in the Bow Street area. These early crime fighters (detectives) became known as the Bow Street Runners, and their success was known to all in London. In 1752, Fielding, an author prior to becoming a magistrate, began publishing *The Covent Garden Journal* to circulate crime news. This literary paper was used as a platform for Fielding's crusade against misery, vice and crime.[3]

Even though Fielding's efforts had immense effects in the Bow Street area, his ideas were not applied throughout London. Crime continued to be a major problem, and society's only weapon against crime was the ineffective constable. Fielding's proposal to have salaried magistrates with a preventive force of paid constables went unheeded.

In 1796, Patrick Colquhoun published *A Treatise on the Police of the*

Metropolis, which detailed the crime problem in and around London. Colquhoun estimated that the losses from various forms of theft, coining, forgery and swindling amounted to £2,000,000. He called for the formation of a large police force to combat crime in London. In 1785, William Pitt introduced a bill resembling Colquhoun's plan in Parliament. Pitt was met with a storm of protest and was forced to withdraw the proposal. The citizenry was adamantly opposed to the formation of any formal police for fear that such a force would be used by the government, or certain elements of the government, to spy on the people, infringe upon liberty, and possibly to aid in the formation of a totalitarian government.[4]

Colquhoun, however, did get a chance to implement some of his ideas. In 1798, a number of West India planters and merchants asked him for suggestions to alleviate the problem of massive thefts from ships and the London docks. Colquhoun developed a plan for a police organization to control the docks. The merchants, with the approval of the government, financed the organization. A river police office was inaugurated with 80 permanent and 1,120 part-time police. The police not only watched and patrolled the docks, they also participated in the unloading of cargo from ships. The experimental police department was a success. Savings as a result of the reduction of thefts was estimated to be £66,000 in the first eight months. The government assumed control of the department in 1800 and operated it until 1829 when it was incorporated into the Metropolitan Police Department.[5]

The Peelian Reform

In 1822, Robert Peel was appointed Home Secretary, and he immediately set out to reform the police. In addition to eventually being responsible for reforming the police, he did make other contributions:

> During Peel's first few years in office, he concerned himself primarily with social reform. First, he consolidated the laws dealing with theft and the destruction of property into one volume. He then did the same thing with all laws dealing with offenses against persons. In England at this time there were more than 200 offenses bearing the death penalty. Peel abolished more than a hundred of these. Benefit of Clergy, where a clergyman could escape punishment for a first offense in certain felonies, was abolished. He made it easier for victims of sexual offenses to get justice by abolishing prior-required embarrassing evidence.[6]

In 1828, Peel appointed a Select Committee to study the police, and on July 27, 1828, they issued their report. The report called for the formation of an Office of Police under the Home Secretary, and all magistrates without bench duty would report to the Home Secretary. All police, constables and watchmen would be incorporated into the Office of Police. London proper was not placed under the structure, which aided in its acceptance by Parliament.[7]

The Bill passed Parliament without serious argument; the most important provision of the Bill was that it made Parliament responsible for finances and administration, eliminating weak, fragmented local control.[8] The Bill also addressed a number of other important areas. For example, there was a section on discipline directed to both the police and the citizenry:

> No policeman on duty could go into a public house except in pursuit of duty. A . . . keeper of any house, shop, room, or other place for the sale of any liquors, whether spirituous or otherwise, who entertained or sold to a policeman could be fined up to five pounds.[9]

Peel appointed Sir Charles Rowan and Sir Richard Mayne as the first Commissioners for the Metropolitan Police. Rowan was selected for his miliatry background, and Mayne, a former Magistrate, was probably selected because of his legal background.[10] One of their first actions was to prepare a book of *General Instructions* delineating the constables' duties and responsibilities.

One of Rowan and Mayne's important contributions was the list of nine principles which guided their department.[11]

1) To prevent crime and disorder, as an alternative to their repression by military force and severity of legal punishment.
2) To recognize always that the power of the police to fulfill their functions and duties is dependent on public approval of their existence, actions and behavior, and on their ability to secure and maintain public respect.
3) To recognize always that to secure and maintain the respect and approval of the public means also the securing of the willing cooperation of the public in the task of securing observance of law.
4) To recognize always that the extent to which cooperation of the public can be secured diminishes proportionately the necessity of the use of physical force and compulsion for achieving police objectives.
5) To seek and preserve public favor, not by pandering to public opinion, but by constantly demonstrating absolutely impartial service to law, in complete independence of policy, and without regard to the justice or injustice of individual laws; by ready offering of individual service and friendship to all members of the public without regard to their wealth or social standing; by ready exercise of courtesy and good humor; and by ready offering of individual sacrifice in protecting and preserving life.

6) To use physical force only when the exercise of persuasion, advice and warning is found to be insufficient to obtain public cooperation to an extent necessary to restore order; and to use only the minimum degree of physical force which is necessary on any particular occasion for achieving a police objective.
7) To maintain at all times a relationship with the public that gives reality to the historic tradition that the police are the public and that the public are the police, the police being only members of the public who are paid to give full-time attention to duties which are incumbent on every citizen, in the interest of community welfare and existence.
8) To recognize always the need for strict adherence to police executive functions, and to refrain from even seeming to usurp the powers of the judiciary or avenging individuals or the State, and of authoritatively judging guilt and punishing the guilty.
9) To recognize always that the test of police efficiency is the absence of crime and disorder, and not the visible evidence of police action in dealing with them.

The principles enumerated by Rowan and Mayne focused on a system of policing where the police were partners with the public. Rowan and Mayne realized that survival of their new police system was dependent upon the public's acceptance. They emphasized cooperation, justice and equality, and crime prevention. The Metropolitan Police represented the first modern police force in history.

The establishment of the police department was not without opposition. Hostility to the new police ranged from brutal murders of the newly-appointed constables to public denunciation by judges, magistrates, cabinet members, the public, and on occasion by King George IV himself.[12] Frequently, the constables were referred to as "Peel's bloody gang" and "blue devils." The police were constantly in fear of their lives, and it was only through the efforts of Rowan and Mayne that the police were able to succeed. They impressed upon the officers to be polite at all times and to use physical force as a last resort. This minimized negative interactions with the public.

Eventually, the police were a success. By June of 1830, the force consisted of 3,314 men. Between 1829 and 1831, 8,000 men had been enrolled, and over 3,000 had been discharged for unfitness, incompetence, or drunkenness.[13] The police brought a reduction of crime, control of riots, and orderliness to London. The police concept was extended to the boroughs in 1835 and to the counties in stages in 1839 and 1856. Gradually, it spread throughout the British Empire.[14]

Early American Police Development

When the colonials arrived in America, they promptly instituted a police system similar to the English system. The Massachusetts Bay Colony installed the office of constable whose duties centered around keeping the peace, raising the hue and cry, controlling drunks and apprehending criminals. Over time his duties were expanded.

> . . . by 1658 they included informing the magistrator of "new comers," taking charge of the Watch and Ward, raising the hue and cry, tallying votes for deputies to the general court, summoning jurymen for duty, bringing accused before the court, bringing before the court men and women not living with their spouses, collecting taxes, and other sundry duties including the hanging of sheepkilling dogs where the owners refused to do so themselves.[15]

As the small colonial settlements developed into cities, night and day watches appeared. In 1631, Boston established a night watch, and in 1643 a burglar watch was established in New Netherlands (New York).[16] In 1700, Philadelphia established a night watch where all citizens were obligated to take their turns.[17] These early watchman systems were not without their problems. As early as 1642 the town government of New Haven proclaimed "It is ordered by the court that, from hence forward, none of the watchmen shall have liberty to sleep during the watch."[18] Many cities experienced difficulty in inducing citizens to take their turns at the watch, and when on duty, many of the watchmen would drink or sleep.

Fosdick analyzed the evolution of the New York watch force, and found the following conditions:

> Its ranks were made up for the most part of men who pursued regular occupations during the day and who added to their incomes by serving the city at night. "Jaded stevedores, teamsters and mechanics" comprised the New York force. No standards except those of a political nature were applied in selection. One Matthew Young was appointed watchman in Boston "in order that he and his children do not become town charges." An investigating committee of the Board of Aldermen in New York made the finding that the incumbents were selected for political opinions and not for personal merit and that the term of service of the incumbent was uncertain and often very brief, depending on the change of political party. Another investigation in 1838 showed the watchmen dismissed from one ward for neglect or drunkenness found service in another.[19]

In 1844 the legislature of New York abolished the watchman system and created a police force. The act established a force of 800 men under the direction of a chief of police, who reported to the mayor. Boston, as did other cities, followed New York's example and established a police force in 1854.[20]

The establishment of the new police forces may have solved some problems, but they also created new ones. From the beginning most of the major municipal departments were embroiled in politics. The *spoils system* more or less dominated the administration of many departments for most of the nineteenth century. The spoils system of local governments caused many departments to deteriorate to nothing more than welfare systems for political cronies. Friends of politicians would be appointed as police captains over lieutenants and sergeants with years of experience. In some jurisdictions, neophyte officers could secure employment only through bribery. Also, if the politics changed, there was generally wholesale firing of police officers, commanders and administrators; the new mayor would appoint his people into the police department to aid him in controlling the city.

The last half of the nineteenth century and the first decade of the twentieth century saw the police become puppets in the politicians' hands. Their primary function was to maintain the status quo. America had failed to learn from England's mistakes. A decentralized police system was allowed to develop which was corrupt and completely inadequate. The American police were, to a degree, a part of the criminal element, rather than being a force which controlled criminality.

Development of Private Security in the United States

Colonists settling in America were confronted by a new and alien land. They brought with them the English system of government and its reliance upon mutual protection and collective responsibility. Although there was a degree of Dutch, Spanish and French influence in the colonies, the preponderance of legal concepts and security practices stemmed from England.

In the New England area, the people depended upon commerce and industry for their livelihood. Thus, many villages and towns were established with most of them utilizing the watch system of their homeland (England) to provide for their security and safety. In the more rural South, the tendency was to develop the county form of government. As in English counties, the sheriff became the principal law enforcement officer in the southern rural areas of America.

As the Atlantic coastal area became more heavily populated, people began to move westward across the country to settle new land. This westward movement was challenged by Indians and other nations not aligned with English interests. Many of these early settlers lost their lives and property to hostile forces who saw the newcomers as trespassers on their land. The fort became a common sight on the American Frontier.

Many of the Frontier areas were without any type of official law enforcement. Thus, the people were mutually dependent upon each other for protection from outlaws and Indians. The wagon train was one such method of protection. Security and safety for the individual was strengthened by traveling with others. It was not an uncommon event to find as many as a hundred wagons traveling together across the western prairie. The 1800's saw the West at its wildest, with murder, gunfights, fraud, landwars, and political graft as the order of the day.

The rapid growth of the American colonies and the westward movement was characterized by increasing needs and pressures for more effective protection. It soon became apparent that the watch system was neither adequate nor efficient enough to match the escalating needs of society. This realization led to the development of public forces in the United States. By 1856 police departments had been formed in New York, Boston, Philadelphia, Detroit, Chicago, Cincinnati, Los Angeles, Dallas, and San Francisco. Although often inefficient and corrupt, they represented a vast improvement over the watch system. Their development, however, did not satisfy or alleviate the growing security needs of private citizens. Thus, the mid-nineteenth century saw the birth of the private security industry.

The Pinkerton Era

Allan Pinkerton is considered the father of private security in the United States. In 1855, Pinkerton, a former Cook County, Illinois deputy sheriff, started the Pinkerton Detective Agency. For more than fifty years his was the only company providing security and investigative services throughout the United States. During the Civil War, Pinkerton agents acted as an intelligence gathering unit for the Union Army, but the primary employer throughout the earlier years of the company was the railroad industry. His detectives and investigators concentrated on catching train robbers and providing other security services for the railroads.

The Pinkerton Agency was successful because public law enforcement agencies were unable to provide adequate protection and security to private citizens and private enterprises. In addition, Pinkerton agents were able to engage in interstate activities unhampered by jurisdictional lines that restricted the public police agencies.

Within a few years of its inception, the Pinkerton Agency was a successful private security enterprise. The foundation was established that the provision of security and protective services could be provided by private enterprise in such a way that both the interests of government and private individuals could be served.

Today, Pinkerton is international in scope, and employs thousands of individuals in a variety of security services and activities.

Other Nineteenth Century Developments

While the English were experiencing the difficulties of establishing a working and fair system of justice and security, American counterparts were confronting similar problems. Between 1800 and 1840, the country experienced a rapid westward expansion and unprecedented changes in the American way of life. The dynamic and expansive nature of American life in issues of statehood, politics, justice, and immigration added new problems and hazards for its participants. In this period alone, immigration more than tripled the national population, introducing a new population with varying personal concepts and cross-cultural concerns.

Security Delivery Services

Stagecoach lines during the westward expansion provided passenger, mail, and courier service throughout the country. Wells Fargo and others were the forerunners of armored car and courier services that are visible today. By 1900, Brink's Inc., begun as a freight and package delivery service in 1859, had a fleet of 85 wagons transporting numerous materials, including payrolls and other valuable goods which could not be shipped by other means.

Electronic Alarms

In 1858, a Bostonian, Edwin Holmes, began the first electronic burglar alarm business. The Holmes Protection Company expanded and adapted to technological advances of the day, and by 1880 the company was monitoring business establishments in Boston, New York, and Philadelphia. Perhaps Holmes' most notable contribution to the development of electronic alarm security was his central-station concept of monitoring various alarm sensors installed at numerous locations from a single location. Holmes' electronic burglar alarm predated the electric light and telephone by some twenty-five years. When the American Telegraph Company (ADT) was formed in 1874, the use of electronic alarm systems spread to most major U.S. cities. By the latter part of the century many companies across the country offered local or district protection.

Railroad Security

As early as 1865, Railway Police Acts were established in many states granting the railroad industry the right to establish a proprietary security force and, in most cases, these forces were given full police powers for the protection of company equipment, rolling stock, and property. By 1860 there were 30,626 miles of railroad in the United States, more than triple the number in existence only 10 years before.[21] With the westward expansion, railway lines were extended into sparsely settled areas that had little or no public protection. By the very nature of their physical construction, trains and railroad properties became a prime target for attack by Indians and well-organized bands of outlaws who robbed passengers, stole cargo, caused train derailments, destroyed property and generally disrupted communications and railway traffic. To combat these problems, watchmen and detectives were hired by the railroads.[22]

The early days of railroad police were characterized by confusion, distrust, and inefficiency. There was no central agency for railroad companies within or without to develop and coordinate standards of employment, job requirements, or public relations. The hiring of individuals was done with little or no regard for their overall ability and background, and a number of undesirable characters found their way into the category of railroad police officer.[23] The measure of a railroad detective or special agent was his ability to handle himself successfully in physical contact with those who preyed upon the railroad. The tact and investigative ability of an agent was most often subordinate to his expertise in handling a sixgun.

By the early 1900's some 14,000 railroad police were employed as investigators and patrolmen.[24] While the railroad police were granted some police powers relative to the protection of railroad operations and properties, they have been, and still are, a proprietary security force with limited law enforcement powers.

The Twentieth Century

By the turn of the century, increased industrialization, immigration, labor organizations and an expanding economy created many conditions that were conducive to the growth of private security. Because public police forces were organized exclusively on a local basis and their operations were limited by local political boundaries, law enforcement beyond restricted jurisdictions was provided only by private security forces. It was not until 1924 that the Federal Bureau of Investigation (F.B.I.) came into being to provide law enforcement on a nationwide, centralized basis. Until this time, private security agencies such as Pinkerton and Burns were the only agencies within the United States with the capability to provide cross-jurisdictional protection of persons and properties.

Prior to and during World War I, the concern for security intensified in

American industry, due not only to urbanization and industrial growth but also to the fear of sabotage and espionage by politically active nationalists. The private security industry experienced a short period of rapid growth and expansion as security personnel supplied by private contractors were utilized to guard the nation's factories, utilities, and transportation systems. The end of the war saw a decrease in security concerns, and the status and quality of security services were lowered.

In the 1930's, plant protection and the provision of corporate security services began to appear. Even though the country was suffering the throes of the depression years, demands for security developed in reaction to labor strikes and unrest in American industry. The Pinkerton Agency alone had informers planted in 93 organizations, many of them holding high union offices.[25]

The beginning of World War II was a tremendous catalyst for the growth of private security services. Almost overnight, thousands of security personnel were employed in the protection of the nation's industries and working forces. The federal government in many instances required contractors to employ comprehensive security measures to protect materials necessary for the war effort from sabotage and espionage. Wartime concern for the prevention and detection of espionage and sabotage brought a federal decision to bring plant watchmen and security personnel into the army as an auxiliary unit of the military police. Before the end of the war, more than 200,000 industrial security personnel were sworn in by the Internal Security Division of the War Department and required to sign an agreement placing them under the Articles of War.[26] As a result of this heightened emphasis and attention to security by the government, private industry, in turn, became more aware of the role that plant security could have in the protection of their assets.

The development and expansion of private and public security services has since evolved from the embryonic stage to include some of the most progressive operational and technological techniques of crime prevention, detection and apprehension in use today. As crime and social problems outgrew the capacity of public law enforcement agencies to provide an umbrella of public services to varying segments of society, there has been a corresponding increase in the role and function of security and protective services. Contemporary security operations have the primary function and responsibility of providing an effective and efficient protection program for the total organization.

There are now more private security personnel than public law enforcement officers in the United States. During the past fifteen years, the growth rate of the private security industry has far surpassed that of public law enforcement. A comparison and projected growth of public sector and private sector protective service workers is shown in figure 1-1.

Figure 1-1

Projected Growth in Protective Service Workers.

	1980	1990	Ten Year Increase
PUBLIC SECTOR			
State and Local			
Police Officers	92,981	108,642	16.8%
Patrolmen	92,972	458,922	16.8%
Sheriffs	22,276	26,601	16.8%
Police Detectives	42,705	49,913	16.8%
	550,504	643,438	16.8%
Parking Enforcement Officers	7,379	8,653	17.3%
Guards and Doorkeepers	25,170	42,428	68.6%
TOTAL STATE & LOCAL	563,053	694,619	23.4%
Federal			
Police Officers	9,179	9,905	7.9%
Police Detectives	20,635	22,267	7.9%
Guards and Doorkeepers	8,987	9,608	7.9%
All Other Workers	1,825	1,969	7.9%
TOTAL FEDERAL	40,356	43,839	7.9%
TOTAL PUBLIC SECTOR	623,409	738,438	18.4%
PRIVATE SECTOR			
Guards			
Proprietary	271,308	369,964	36.4%
Contract	341,102	443,594	30.0%
Store Detectives	18,279	27,365	49.7%
Fitting Room Checkers	8,864	11,790	33.0%
Security Checkers	230	260	13.0%
Railroad Police	2,165	1,944	[10.2%]
All Other Workers	2,395	3,976	[7.9%]
TOTAL PRIVATE SECTOR	644,343	858,893	33.3%

Source: National Industry-Occupation Matrix, 1980-1990, Bureau of Labor Statistics, 1982, p. 108.

Note that several identifiable classes of security workers are missing from the employment estimates, including security consultants, armored car personnel, and alarm company personnel.

It is likely that the number of people employed in the private security industry is larger than the BLS estimate. In 1976, the Private Security Task Force estimated that as many as one million persons might be employed in security with the organizations represented by the membership of the American Society for Industrial Security (ASIS). The Hallcrest Report on private security and police in America estimated private sector security employment in 1982 at 1,100,000, excluding government security workers.[27] In any case, it is evident that the field of private security is growing and involves a broad range of position classifications and occupational groups of persons engaged in the tasks of guarding, investigating, surveilling, transporting, inspecting, installing, consulting, monitoring and servicing protective products.

Factors of Growth

Guard Service Expansion

Private, contractual guard service firms such as Pinkerton, Burns, Wackenhut, and Guardsmark have achieved enormous growth during the past few years. Proprietary security services (companies providing their own security guards) while not as visible as the contractual services, experienced equal if not greater growth. The activities and functions of guard service operations are essentially the same whether they are contractual or proprietary. Protection of lives and property, crime prevention, access control, order maintenance, and information security are basic functions in providing facility and personnel protection. As shown in figure 1-1, proprietary and contract guards account for a large percentage of the total personnel employed in the private sector of security. Today, contract security guards and proprietary security personnel can be found in almost every segment of commerce, manufacturing, institutional services, and government.

Technological Advances

Technology has played an important role in the growth of the private security industry. Even the lock, one of the oldest security devices, has been subject to drastic changes: combination locks, time locks, electronic locks and access-control systems that incorporate the advanced technology of television and minicomputers.

Technology that has improved the quality of television, radios, communications and other areas of electronics has been adapted and assimilated into electronic security devices and systems. The progression from vacuum tubes to transistors to integrated circuit technology has played a major role in the growth of the security industry. Today, electronic security products and services comprise a sizable portion of the security market.

Rising Crime Rate

Without question, the rising crime rate has contributed to the growth of security in the United States. Crimes against persons and property in recent years have increased to the point that many live in constant fear of being victimized on the street, at work, and even in their homes. Guard services and electronic alarm devices are now seen as ways to provide measures of security that would deter or detect would-be criminals from stealing property or causing personal injury.

In addition to the age-old crimes of murder, theft, rape, assault, etc., new crime techniques are being utilized to perpetrate criminal acts. Computer crimes, credit card fraud, and other forms of white-collar crime are relatively new and in many cases are beyond the scope of effective investigation by public law enforcement agencies. These types of crime have created a need for a new kind of security person capable of developing and employing highly sophisticated security measures to deter, detect, or deny criminal access. The complexity of today's business environment can sometimes be staggering, yet not so complex that an enterprising employee cannot find ways to embezzle, defraud, or otherwise steal from his/her employer. Recent studies have indicated that perhaps the employee represents more of a threat to the employer than all the robbers, burglars, and outside thieves combined. Clearly, much of the growth of security has been generated by the threat from within.

Other crimes, while not new, have become more prevalent in recent years. Principal among these are terroristic acts including skyjacking, political kidnapping, bombing, assassination, and the holding of hostages for ransom. Airports have been the scene of bombings, mass murders, and holding of hostages. Government buildings have been attacked by suicide squads. Crowded shopping centers have experienced bombings where large numbers of shoppers were killed and injured. Diplomats, vacationers, and businessmen have been assassinated, kidnapped and held for ransom in many parts of the world. These situations, whether politically or criminally motivated, have created a need for security personnel with the knowledge, professional characteristics and skills to deal with the terrorist threat. In many instances, private companies, public institutions, and governments have chosen to initiate additional and often elaborate security measures to thwart such criminal acts.

Government Regulation

The Federal Bank Protection Act of 1968 mandated increased security measures and equipment for federal banks. Since January 1973, the Federal Aviation Administration has required screening of air passengers and their carry-on baggage. Many Department of Defense contractors must comply with stringent security guidelines regarding the hiring of employees, facility access controls, procedural operations, and security of information. Government regulations created a corresponding demand for security personnel and products to satisfy the new standards.

Professional Growth

In 1955 the American Society for Industrial Security was formed. Its past and current membership is made up of security practitioners whose purpose is to advance and enhance the security profession. ASIS serves as a major spokesman for the security industry and, in recent years, has focused on the need for advanced research in loss prevention, crime reduction, and advanced security education. Additionally, ASIS has a Certified Protection Professional (CPP) program with the major objective of fostering professionalism in the field of security. Candidates must meet experience and educational requirements and pass an in-depth one-day examination before receiving their Certified Protection Professional certification. Other security organizations include the National Council of Investigation and Security Services, National Locksmith Association, and the National Burglar and Fire Alarm Association.

Security has become a complex and diverse field. The traditional concept of the aging, less-than-active night watchman will no longer suffice. Instead, there is emerging a new security person, well trained, more highly educated and better able to satisfy the growing intricacies of the security profession.

Conclusion

Security in America can find its historical roots in the scope and range of events that shaped early England. English concepts of law, justice, social structure and security evolved and were gradually assimilated into early American life. One merely has to look at the historical occurrences and societal evolutions of England to discover why and how certain changes occurred in America.

Until 1776 the dominance and influence of England over the American colonies was indisputable, and even afterwards many practices and traditions of the colonist were English born. America's developing law enforcement and justice system was for the most part patterned after the English system. As the nation grew so did its problems. Efforts to begin municipal police departments were fraught with problems of political favoritism and corruption. The westward expansion was characterized by Indian attacks, range wars, outlaw gangs, and sparse law enforcement. At every turn, new problems requiring new solutions were confronted by our forefathers.

Security in the private sector began and grew in direct relationship to the needs of American society. The Civil War, the development of an extensive railway system, World Wars I and II, technological advances, a rising crime rate, government regulations, and other factors have in the passage of time combined together to initiate and nurture a significant and growing demand for security products and services.

Discussion Questions

1. In what ways did people in ancient times provide security for themselves and their societies?
2. What influence did the Anglo-Saxons have on the development of early England?
3. What changes did William, Duke of Normandy, initiate in Eleventh Century England?
4. What modern day security practices can be traced to the Statute of Winchester?
5. What contributions did Henry Fielding, Patrick Colquhoun and Robert Peel make to the development of security and law enforcement?
6. What were some of the major problems encountered by law enforcement efforts in Nineteenth Century America?
7. Discuss the origin and development of railroad security.
8. Discuss the factors, positive and negative, that have influenced the growth of security in the Twentieth Century.

Notes

1. Patrick Pringle, *Hue and Cry* (Suffolk, England: Richard Clay and Company), p. 11-19.
2. Charles Reith, *The Blind Eye of History* (Reprint Edition, Montclair, N. J.: Patterson Smith, 1975), p. 135.
3. Thomas R. Phelps, Charles R. Swanson, Jr., and Kenneth R. Evans, *Introduction to Criminal Justice* (Santa Monica, California: Goodyear Publishing Company, Inc., 1979), p. 43.

4. Reith, p. 138.
5. Reith, p. 138.
6. William H. Hewitt, *British Police Adminsitration* (Springfield, Illinois: Charles C. Thomas, 1965), pp. 23-24.
7. Hewitt, p. 24.
8. Hewitt, pp. 23-24.
9. J. L. Lyman, "The Metropolitan Police Act of 1829; An Analysis of Certain Events Influencing the Passage and Character of the Metropolitan Police Act in England," in *Issues in Law Enforcement*, edited by G. Killinger and P. Cromwell (Boston, Mass.: Holbrook Press, 1975), p. 31.
10. Hewitt, p. 30.
11. Reith, pp. 154-166.
12. Reith, p. 152.
13. Lyman, p. 37.
14. Reith, p. 169.
15. Phelps, Swanson, and Evans, p. 41.
16. William J. Bopp and Donald O. Schultz, *Principles of American Law Enforcement and Criminal Justice* (Springfield, Ill.: Charles C. Thomas, 1972), pp. 17-18.
17. Raymond B. Fosdick, *American Police Systems* (Reprint Edition, Montclair, N. J.: Patterson Smith, 1969), p. 59.
18. Fosdick, p. 60.
19. Fosdick, pp. 61-62.
20. *Task Force Report: Private Security Standards and Goals* (Cincinnati, Ohio: Anderson Publishing Company, 1977), p. 30.
21. H. S. Dewhurst, *The Railroad Police* (Springfield, Ill.: Charles C. Thomas, 1955), p. 5.
22. Dewhurst, p. 6.
23. Dewhurst, p. 7.
24. Richard Post and Arthur A. Kingsbury, *Security Administration: An Introduction* (Springfield, Ill.: Charles C. Thomas, 1970), p. 15.
25. Henry S. Ursic and Leroy E. Pagano, *Security Management Systems* (Springfield, Ill.: Charles C. Thomas, 1974), p. 19.
26. Ursic and Pagano, p. 20.
27. William C. Cunningham and Todd H. Taylor, *The Hallcrest Report: Private Security and Police In America* (Portland, OR: Chancellor Press, 1985), p. 113.

Chapter 2
Current Status of Security Services

During the past three decades administrators in many areas of our social, government and economic systems have recognized that all problems of disruptions, thefts, vandalism, assaults, and hijackings could not be solved through the traditional public police agencies. An expanding economy and an increased crime rate meant that the police agencies did not have the manpower to fully protect large businesses and industries, hospitals, etc. These private sector needs for security of property and persons created or expanded guard services, protective alarm services, armored car services, private investigative services, locksmith services and security consultant services (figure 2-1). The cost and inconveniences

Figure 2-1
Private Security Service Overview.

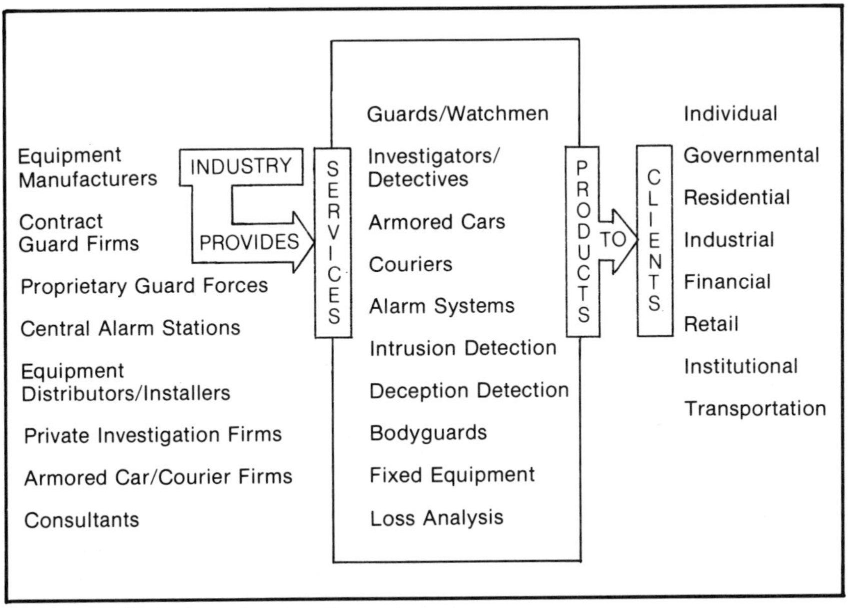

Equipment Manufacturers
Contract Guard Firms
Proprietary Guard Forces
Central Alarm Stations
Equipment Distributors/Installers
Private Investigation Firms
Armored Car/Courier Firms
Consultants

INDUSTRY
PROVIDES
SERVICES

Guards/Watchmen
Investigators/Detectives
Armored Cars
Couriers
Alarm Systems
Intrusion Detection
Deception Detection
Bodyguards
Fixed Equipment
Loss Analysis

PRODUCTS
TO
CLIENTS

Individual
Governmental
Residential
Industrial
Financial
Retail
Institutional
Transportation

Source: Private Security: Task Force Report.

associated with numerous acts of unlawful conduct caused managers in such areas as transportation, commerce, health care, retailing, industry, government and schools to look toward alternative means of protection. Even though private security had been established for decades in major companies and industries, security was foreign to most quasi-governmental and/or public institutions. Losses associated with a variety of factors caused more areas of private enterprise to look to security for assistance. The current status of most major areas of security is discussed below.

Security Services

There are thousands of people employed in a variety of private (nongovernmental) security units. Some of these people perform limited security duties, such as do some receptionists and night watchmen assigned to often perfunctory positions of responsibility and authority. Others work in highly specialized fields, such as arson investigation and alarm services. Regardless of their level of responsibility and expertise, these people are involved in the overall protective services of our nation and have a direct and important bearing on crime prevention and crime reduction. This, in turn, requires that private security units work very closely with the formal governmental units of law enforcement (figure 2-2).

Guard Services

To the general public, the uniformed security guard seen at retail stores, industrial plants, office complexes, banks, hospitals, sports complexes, and governmental facilities is the most visible part of the private security industry. Some wear the distinctive insignia of that organization, while others wear the insignia of a private, contractual guard firm. There are, then, two distinct types of guard services.

Contract Guard Services

This type of guard service is purchased from a firm outside the organization, generally for a rate per guard hour. The rate per guard hour that the contract security firm receives covers the costs of the guard's salary, fringe benefits, worker compensation, liability insurance, office expense, telephone services, travel and supervisory salaries. The guard personnel are employees of the contracting firm. Their duties and responsibilities, whether highly technical or mundane, are defined by the contract and administered by the contracting agency. Management and/or supervisory personnel of the contracting firm are usually consulted for ad-

Figure 2-2
Protective Services and Products
Projection of Revenues and Value of Shipments to 1995.
($millions)

Categories	**1985[1]**	**Projected 1995[2]**
Revenues		
Protective Services		
Guard & Investigations	4712.0	5654.4
Central Station Alarm Monitoring	1225.0	1470.0
Armored Car	487.5	585.0
Protective Services **Total**	6424.5	7709.4
Value of Shipments		
Deterrent Equipment		
Fixed Security Equipment (safes, vaults, cabinets, etc.)	316.1	379.3
Locking Devices	1160.7	1392.8
Electronic Access Control	31.4	37.6
Security Lighting	530.0	636.0
Data Encryption Devices	20.1	24.1
Security Fencing	3729.0	4474.8
Deterrent Equipment **Total**	5787.3	6944.6
Monitoring and Detection Equipment		
Intrusion Alarm Systems		
Local and Proprietary	303.4	364.0
Central Station	134.4	161.2
Direct Connect	87.5	105.0
Hold-up Systems	1.2	1.4
Intrusion Alarm **Total**	526.5	631.6
Closed Circuit T.V.	124.2	149.0
Electronic Article Surveillance	97.3	116.7
Monitoring and Detection **Total**	748.0	897.3
Fire Detection and Control Equipment		
Fire Alarm Systems		
Local and Proprietary	193.0	231.6
Central Station	105.5	126.6
Direct Connect	46.1	55.3
Fire Alarm Systems **Total**	344.6	413.5
Fire Extinguishers	400.0	480.0
Smoke Detection	211.3	253.5
Automatic Sprinkler Systems	1000.0	1200.0
Fire Detection & Control Equipment **Total**	1955.9	2347.0
Protective Services & Products Total		**$17,893.3**

1. *Hallcrest Report: Private Security and Police In America,* William C. Cunningham and Todd H. Taylor, Chancellor Press.
2. Projections based on: *Occupational Projections and Training Data,* U.S. Department of Labor, Bureau of Labor Statistics, 1984.

vice and assistance before any new program is implemented. Most new security programs are implemented by the personnel from the contracting firm. Such contracting agencies may be very large international organizations, such as Pinkerton or Burns, or they might be very small and operate in a very limited geographical area.

In recent years, contract guard services have grown and changed in response to economic and technological developments, reductions in the funding of public services, and changing client demands. Developments have led to an industry-wide focus on several important issues: 1) the trade-off between cost and quality service, 2) liability and insurance, 3) the delivery of quality service, 4) client expectations of quality and protection, 5) training, 6) guard turnover, and 7) government regulations.

Once wages, benefits, and fees are agreed upon and the contract is signed, contract guard firms must take steps to ensure that the quality service they advertised will actually be delivered. Adequate funding to allow good pay and benefits is only the first step. Good hiring practices, training programs, close supervision, and a sensible management structure are also important.

The issue of liability and insurance is generally recognized as a serious problem for the security field. Liability issues are closely aligned to employee turnover, training, and quality of employees. As more security personnel are involved in questionable practices and more suits are filed, and the courts award plaintiffs large settlements, contract security firms find their insurance rates rapidly escalating. Most insurance companies are becoming increasingly reluctant to insure high-risk contract security firms.

Government regulations, especially state statutes, will provide the needed impetus to establish standards for employment and training. Such standards should dictate a balance between contract and proprietary guard services and armed/unarmed personnel.

Proprietary Guard Services

This type of guard service is often referred to as *in-house* security, that is, the security personnel are employees of the organization being protected. Salaries and other benefits are paid directly to the employee in contrast to the way that a contracting firm's personnel are paid. Duties and responsibilities are defined and controlled by the organization, not be an outside agency. Managers and supervisors of proprietary security personnel, being a part of management, are often involved in the corporate decision-making processes of planning and implementing programs in the areas of security and safety. Many such in-house security forces are referred to as *plant protection* units, because their duties include accident prevention and investigation, complicated control of traffic and pedestrian movement, clearance and escort of non-employees, fire prevention and protection, as well as basic law enforcement within the facility.

Protective Alarm Services

Many commercial enterprises, industrial operations, institutional facilities, and private homes utilize a wide variety of alarm systems and services. Intrusion detection alarms, robbery alert alarms, fire detection systems, equipment for manufacturing process alarms, and other specialized alarms, often technically sophisticated, are installed, maintained, and monitored by many agencies throughout the country. These alarm systems can be either proprietary or contractual, that is, they may be installed, maintained, and monitored by the user or by a contract agency. Regardless of the type of protection or the ownership, there are three basic types of alarm systems:

1) The Local Alarm System – This type is designed to sound an on-premise alarm, such as a horn, bell, or siren. It serves to scare would-be intruders away and to alert persons in or near the premises.
2) Direct Police Connect – This type of system sounds no audible alarm on the premises. Instead, a silent signal is sent to the police department, via telephone lines. The police dispatcher then alerts the police unit nearest to that location.
3) Central Alarm System – This type of alarm system also utilizes the silent alarm signal, but instead of being monitored by the police the signal terminates at a remote central alarm station. Central alarm stations may be either contractual or proprietary. Response to alarms may be made by contract guards, proprietary guards, or the local police who were notified by the central station attendant.

Armored Car Service

Some of the private units of protection specialize in guarding and transporting cash, securities, gold, jewelry, or other valuables. These security operations provide a distinct service to financial institutions, commercial operations and others who must transport valuables from one location to another. One of the oldest of such agencies is Brink's, Inc., established in 1889.

Private Investigative Service

Private investigative agencies, from one-man local operations to multi-employee international operations, offer their services to private citizens, attorneys, and commercial and industrial enterprises. Private persons who desire discreet investigation of members of their family, potential family members, or perhaps even criminal situations or unsolved crimes where the client has a personal interest, will often hire private investigators. Attorneys who desire information relative to clients, witnesses, jurors, suspects, or opposing parties often utilize the services of a private investigator. Industrial and commercial concerns that desire credit information, background checks on potential and current employees, or information on competitors employ private investigative services quite extensively.

Locksmith Services

Locksmiths provide a distinct and often very critical function in the security industry. Choosing locking devices and keying systems to fit a particular situation can be best accomplished by the trained locksmith. While locksmiths often can provide a needed security service, their skills are often underutilized. Most locks are manufactured by large, national companies and distributed by various types of retailers and suppliers, and so are more often purchased through retailers and installed by general contractors or in-house maintenance personnel. Such persons very often do not have the expertise and knowledge of security possessed by the professional locksmith.

Security Consultant Service

Private security consultants are somewhat of a new addition to the security industry. These individuals usually have years of experience and can provide valuable assistance to industrial, commercial, or institutional clients who desire outside assistance on security-related problems.

The security services described above do not represent the full array of security services. There are numerous other private security services available, including special security patrols, insurance investigation, and polygraph examination. Of course, a single private security agency may offer a variety of security services to clients.

Specialized Areas of Security

To place the area of security in proper perspective one must recognize that it can be either a service provided by a private agency on a contractual basis, or an integral, operating component of the organization itself. In either case, security cannot function independently of the organization it serves.

In order to get a comprehensive and distinct profile of the application of security services one must look to the whole range of businesses, industries, and organizations where it fulfills a vital function. The type of security service and the degree of application is dependent on the specific needs and problems of the entity being served.

Transportation

Every year, millions of passengers and billions of tons of cargo are processed by the various transportation agencies. Each of the component parts of the transportation system have both common and unique security problems.

Most governmental or quasi-governmental facilities (e.g., airport authorities, port authorities and some mass transit agencies) operate from a different legal position than most facilities owned and operated by private enterprise. For example, security employees of airport authorities, port authorities and some mass transit agencies generally have the same powers of arrest as a regular police officer of that jurisdiction. Proprietary security officers and employees of a contract security firm doing business as a private enterprise venture do not have the same power of arrest as a police officer of a particular jurisdiction.

Generally, unless deputized, commissioned, or provided for by an ordinance or state statute, private security personnel possess no greater legal powers than any other private citizen. However, due to the position occupied by security officers, they have a much greater opportunity to use their citizens' powers than does an ordinary citizen. This justifies considerable attention and training in the laws of local jurisdictions.

Airports and Airlines

The most critical and threatening challenge to safety and security of airlines developed in the 1960's and is known as skyjacking. The number of skyjackings reached its zenith in 1968 and 1969 with most commandeered flights forced to land in Cuba. Monetary gain and political reasons were the predominant motives for skyjacking. One of the most famous skyjacking cases involved Dan Cooper who on November 24, 1971, hijacked a plane after boarding in Portland, Oregon, for a flight to Seattle, Washington. Money and parachutes were demanded and given to him and he bailed out of the plane somewhere over isolated, mountainous terrain between Seattle and Reno, Nevada, never to be heard from again.

Responding to the hue and cry from the public, the Federal Aviation Administration assisted in developing a psychological profile of a terrorist. Subsequently, metal detectors were put into use at airports to screen selected persons based on the established profile. The profile system was only partially effective, and as a result, the sky marshal system went into effect in 1970. Like the profile system, the sky marshal program was only partially effective. Skyjackings still occurred and one was undertaken when both a sky marshal and an FBI agent were aboard the plane. This embarrassing event led to the conclusion that skyjacking could best be controlled from the ground.

With the concept of concentration of security on the ground for airline safety, new rules for airlines and airports were developed: Part 107 of the Federal Aviation Regulations (FAR) for airports, and Section 538 of FAR, Part 121 for scheduled carriers. Airports would be responsible for controlling air operational areas, and carriers would be responsible for preventing sabotage devices and weapons from being carried aboard airliners. Furthermore, it was intended that passengers' baggage would be checked. Since airports and carriers were required to design their own plans, there were many different approaches and arguments about the degree of responsibility for security placed upon airports and that which was required of carriers. Most of the problems were solved in 1976 by the Security Committee of the Air Transportation Association which developed a single, Standard Security Program.

The Anti-Hijacking Act of 1974 and the Air Transportation Security Act of 1974 implemented an international attitude toward skyjacking as expressed by The Hague Convention for the suppression of unlawful seizure of aircraft. Within the scope of international affairs the President has the right to suspend summarily, without notice or hearing, for an indeterminate period of time, the right of any carrier to engage in air transportation between the United States and any nation that permits its lands to be used by terrorist organizations or that promotes or engages in air piracy. Moreover, a secondary boycott could be imposed by the President which would suspend air commerce between the United States and any foreign nation that maintains air service between itself and an offending country.

The Secretary of Transportation was given the power, subject to approval by the Secretary of State, to bring sanctions against airlines which fail to maintain minimum standards set by the Convention on International Civil Aviation.

In 1975, FAR 129, "Foreign Air Corridor Security Program," required all foreign air carriers landing in and departing from the United States to screen all passengers and carry-on luggage. It also provided for the search of planes threatened by terrorists, and prohibited the carrying of weapons in cabins.

Among other provisions of public law, *in flight* was defined to be from the moment all external doors are closed following embarkation until the moment all external doors are opened for disembarking. This definition was necessary to clarify jurisdiction over suspects apprehended during several stages of travel by air. It was also made an offense to attempt in any way to carry a weapon aboard a plane. Under these provisions, the FAA would have exclusive responsibility for the direction of law enforcement activities affecting the safety of persons in flight. Another deterrent was provided when the death penalty was authorized for conviction of a skyjacking that caused a death.

International terrorism is a hazard that faces airlines and airports in the 1980's. It is a relatively easy and inexpensive way for radical dissident groups to strike at the civilized nations of the western world.

Terrorism and Travel. We have experienced the savagery of international terrorists during this decade — an elderly man shot in the head and thrown overboard in his wheelchair — a mother and infant daughter blown out of an airliner by a terrorist bomb — 130 people gunned down while waiting in line at air terminals in Vienna and Rome. The goal of international terrorists is difficult to discern, because in many cases, its agents are fanatics with obscure political agendas. It can be determined that they seek to drive western influence from the developing world, and that they tend to favor despotic political systems over our tradition of plural representative democracy.

Currently, the most active agents of terrorism seem to be groups spawned in the Middle East, but the problem is by no means limited to that region. Countries in South and Central America, Europe, Asia, and Africa have been afflicted by forms of warfare, perpetrated by groups as diverse as the Irish Republican Army and the Shiite Army of God.

American citizens are principal targets of terrorists. Forty percent of all terrorist attacks have been directed at U.S. citizens or installations abroad. Incidents within the United States have been decreasing — together representing less than one percent of the worldwide total. The effective work of the FBI, tighter controls at U.S. points of entry and the natural American aversion to foreign inspired violence have had an effect. The Omnibus Anti-Terrorism Act of 1986 offers up to one million dollars for the apprehension of persons who kill or take Americans hostage, up to $500,000 for information helping to thwart a terrorist plan, upgrades security for U.S. installations at home and abroad, bans the export of American technology to countries which support terrorists and provides compensation to the families of U.S. service personnel and other Americans who fall victim to terrorists.

Airports and airlines cannot provide all of the security that an individual traveler might require, especially if the person is traveling abroad. The following are a few suggestions for safe travel at all locations, not just while at an airport or aboard an airplane:

1) Move directly to the gate – avoid lingering in the airport lobby or other uncontrolled areas. Have your ticket and boarding pass with you when you arrive at the terminal and move directly from luggage check-in to the gate. There you will be in the one part of the terminal where passengers and luggage have been screened and X-rayed.
2) Avoid fancy luggage and dress – in the event of hijacking, expensive luggage and dress tend to label one as a "rich American." Try to avoid appearances that are out of the ordinary.
3) Avoid the aisle seat – hijacking victims seated near the aisle have fared worse than those near the window.
4) Avoid first class – first class passengers are seen as being wealthy Americans and terrorists most often set up their headquarters in the first class compartment.
5) Car rentals – rent inexpensive local cars to remain as inconspicuous as possible.
6) Restaurants – avoid sitting near front windows, which are the most vulnerable to bombing attacks. Avoid restaurants which specialize in traditionally American food or are known to be frequented by Americans.
7) Identification cards – official U.S. government ID cards or military service ID should be left at home or stored in your luggage. Terrorists more frequently prey on those holding official identification.
8) Contact U.S. Embassies and Consulates – in cases of emergency, you should contact the nearest American embassy or consulate for assistance.

Air cargo thefts, passenger checks, skyjacking and terrorist acts are the major security problems faced by airports and airlines. Most airport authorities have proprietary security forces to perform security functions associated with the properties of the airport. Passenger checks, however, are generally performed by employees of contract security firms. A problem unique to airports and airlines is the potential for large-scale disaster. Skyjacking, terrorist acts, and plane crashes are examples of events for which security personnel must be prepared.

Railroads

The railroad police are perhaps the oldest and best organized segment of the private security industry. The railroads pay their security personnel salaries and fringe benefits comparable with other railroad employees, and so they are able to attract and keep an excellent security force. Cargo thefts, vandalism and thefts of metals are the industry's major security problems.

Maritime

Many large cities have a maritime authority which operates under governmental or quasi-governmental authority. Proprietary security forces dominate the maritime field, though the individual companies who lease facilities from the maritime authority may use a combination of proprietary and contract security services. Cargo theft is the major security problem associated with the maritime industry.

Leisure cruise lines are facing a new crime threat in terrorism. With large numbers of travelers on a single vessel, terrorists are beginning to view cruise lines as another means of satisfying their demands. This is exemplified by the hijacking of the Achille Lauro, an Italian cruise ship, by force while in port at Alexandria, Egypt on October 7, 1985, by four terrorists associated with the Palestine Liberation Front. This was the first instance of terrorists commandeering a pleasure vessel.

Trucking

Most of the materials and goods transported in the United States are transported by common carriers, rather than company-owned transportation fleets. Most materials and goods are transported by truck. Trucking firms rely on a small proprietary security force to deal with major thefts and utilize the services of a contract security firm at fixed locations, such as terminals or distrubition centers. Cargo theft, in the form of employee theft or hijacking, is the trucking industry's major security problem.

Transit Authorities

Transit authorities, such as the Chicago Transit Authority, are generally quasi-governmental agencies who are financially supported by both public funds and the revenues from passenger services. One of the primary financial concerns of mass transit systems is the loss of revenues as a result of fear of crime. Robberies and vandalism are two primary security problems. Since a large portion of the security work is investigation and report writing, a proprietary security force is generally utilized by transit authorities. Hardware and architectural design are the latest improvements to be tried by transit systems. Transit authorities are seeking to increase visibility in passenger waiting areas, to reduce patron waiting time and to provide quick detection and response to criminal incidents through better utilization of security officers, hardware and architectural design.

Commerce

Commercial facilities have security problems which are not easily solved by current police practices. Public law enforcement agencies have neither the manpower nor the capability to provide security for the vast number of banks, hotels/motels, and other commercial enterprises in the United States. These businesses, whether isolated or part of a commercial complex, cater to the general public and must encourage a feeling of openness and availability if they are to remain competitive. Yet at the same time commercial enterprises must project an image of solidarity and security. Commercial facilities must provide for protection against an array of criminal activities from the simple theft of items from a motel room to the multi-million dollar computer embezzlement.

Hotel-Motel

Hotel-motel managers must provide their customers with a safe and secure environment. The security problems associated with the hotel-motel industry are for the most part related to crime in the surrounding area. Major security problems associated with the hotel-motel industry are thefts from the automobiles of guests while parked in a parking lot, thefts from the rooms of guests, and vandalism to the property of both the facility and guests. The main concerns here are the density of population, the traffic pattern of both vehicular and pedestrian traffic, the crime rate and the type of crime that prevails in the area, and any other features of the location that may make the hotel more vulnerable to adverse incidents. Such other features are the kinds of businesses in the area, the security protection maintained by those businesses, and the availability of public services nearby that lessen the probability of man-made disaster or criminal trespasses. Of particular importance is the presence or proximity of police, i.e., are streets patrolled by walking patrols or motorized patrols or both? What is the response time of the public police?

Size can be considered in length, width and height. Also, it can be measured by the number of guest rooms, ancillary facilities, parking lots, swimming pools, bars, restaurants, theaters, conference rooms, offices and shops. Generally, the larger the building, the more numerous the guest rooms and other facilities that complicate the security mission. Yet the design of the buildings and utilization of the grounds can offset some of the disadvantages of size. If the building is a high rise with the proper number of entrances and exits which can be controlled consistent with safety, security personnel can be used with economy and efficiency. If the personnel that staff the desks, render personal services, and operate shops and entertainment activities are used as eyes and ears to enhance safety and security, size becomes a relative matter.

The proper choice and use of hardware and anti-intrusion devices further reduces the threat attributed to size alone. Scheduling of security personnel during shifts when they are most needed is important in reducing vulnerability. Most trespassers intrude, and most crimes are committed at night. During the late hours or early morning hours there is less activity in the establishment and correspondingly fewer people to observe intruders. This situation may impact upon the deployment of security personnel and the assignments and locations of other members of the staff.

Not only should security and management consider the obvious hazards that have been pointed out by experience and documentation, but they must look for likely hazards that have not yet manifested themselves. A lesson was learned from a deadly event that occurred in New Orleans, Louisiana in 1973 at the downtown Howard Johnson's Hotel when a demented or depraved trespasser walked into the hotel and killed seven people and wounded a score of others. During the rampage of crime, he committed both murder and arson until he was killed by the police. Security management must plan for the shift of street crime from outside the hotel premises to its grounds and buildings.

Security services are generally provided by a proprietary staff supplemented by employees of a contract security firm. The security manager is the decision-maker and the contract employee performs the patrol-watchman role.

Appendix B includes a hotel/motel security survey or audit instrument.

Office Buildings

Large commercial office buildings typically have elevator banks in the lobby area and security personnel can monitor most of the pedestrian traffic from one central location. Closed circuit television systems are commonplace in office buildings. Security is generally provided by contract security personnel, whose duties include monitoring the CCTV and making regular security checks throughout the building. The major security problems for commercial office buildings are after-hours burglaries and thefts.

Financial Institutions

Financial institutions are faced with security problems quite different from those of other commerical enterprises. In contrast to the many indirect losses sustained by other businesses, most losses to financial institutions are direct financial losses. During the 1960's, financial institutions grew at a rapid rate. Significant increases in robberies, larcenies, and burglaries during that time led Congress to pass the Bank Protection Act of 1968. This Act provided that federally-insured banks and savings and loan associations and, later, federally-insured credit unions: (1) designate someone to be a security officer, (2) cooperate with and seek security advice from various law enforcement agencies, and (3) develop comprehensive security programs and implement protection measures.

The FBI estimates that eight times as much money was stolen by bank officials and other insiders in one recent year as was extracted by bank robbers and burglars. Criminal misconduct by financial institution insiders was a major factor in about one-half of all bank failures in recent years and the situation isn't helped by the punishment meted out: the small number of insiders who are convicted spend less than a year in prison.

Failure to have adequate photographic or video taping equipment available and in place in case of a robbery is a major concern of the FBI. Without a usable photograph or video tape to aid in identification, it makes it virtually impossible to conduct a speedy investigation.

Fraudulent use of bank credit cards and computer crimes are now the responsibility of the FBI. Major losses from both credit card schemes and computer crimes caused Congress to enact federal legislation mandating that such crimes be under the investigative authority of the FBI. Counterfeiting of bank notes are under the investigative authority of the U.S. Secret Service, an agency of the Department of Treasury. Even though many crimes committed against banks are handled by local law enforcement and the FBI, most financial institutions have a security director for decision-making and utilize contract security services to supplement the work of the security director.

Health Care Institutions

Many security problems are unique to the health care industry. A hospital, for example, must remain open to admit the sick and injured, to allow patients to have visitors, and to carry on the normal activities that are required in caring for those who are unable to care for themselves. Because of this openness, adequate physical security, especially access control, is difficult to attain.

Hospital Security

Hospital security has grown many times over in the last three decades. The socio-economic changes in this period affected security at hospitals, particularly in inner city locations. The decline of the inner cities has forced hospital administrators to deal with the problems of security. In the past, hospital administrators were not skilled in the area of security, nor did they want to become skilled. In most hospitals, the security duties were given to an assistant administrator, who also was not skilled in security matters. Recognizing the need for a professional organization to study the problems of hospital security, the International Association of Hospital Security was founded in 1968. Since the inception of IAHS, more emphasis has been placed on having a full-time person who is directly responsible for security services.

IAHS has a certification program for individual members of the hospital security force. Certification is available for security officers, security supervisors and the security director. IAHS also has a certification program for trainers, and offers such programs nationwide on a yearly basis. Most directors of security at hospitals believe that IAHS plays a greater role than the Joint Commission on Accreditation of Hospitals (JCAH) in providing professional security personnel and services due to its emphasis on training. Like all other areas which need security, training or the lack of training is one of the major problems in hospital security, and IAHS is assisting in alleviating this problem.

Another pressure on hospital management to meet certain security standards is the Joint Commission on Accreditation of Hospitals (JCAH) housed in Chicago. It is composed of persons appointed from the American College of Physicians, American College of Surgeons, the American Hospital Association, and the American Medical Association. Hospitals must voluntarily seek accreditation, and those that do must agree to a security survey or audit which includes prescribed standards of performance and operation. A prescriptive accreditation manual is published and updated annually.

One of the major problems facing hospital security is internal, employee theft. Some security personnel believe that it is the source of the largest single dollar loss occurring in most hospitals. Security has the additional responsibility for serving the public and minimizing the chance that any person coming on the premises will suffer serious injury. There are certain standards that must be met to comply with laws, regulations, and codes, particularly in the realm of fire protection and safety.

Other vulnerable areas are receiving areas, cash registers, food services, package and parcel inspection, removal of bodies, lost and found, utilities, accesses, employer locker areas, and parking and traffic control. Areas of vulnerability which overlap with several of the areas listed above and need special attention are: maternity wards, emergency rooms, and visitors and visitor control.

Maternity wards. Footprinting of infants at birth is essential to establishing subsequent identification of newborns. Infant kidnapping and switching of babies by design or accident is of foremost concern in this area. Nurseries must be maintained as a secure area. The maternity ward itself should be strictly guarded with hardware or people. Infants should be in the nursery during visiting hours. Nursery personnel should wear distinct uniforms. A mother must know that persons wearing the distinct uniform will take care of her child. These uniforms must be tightly controlled. When mother and infant are discharged, a staff member should accompany them to the point of discharge and stay with them until they depart.

An example of a maternity ward case occurred on December 11, 1985, when a three-day-old baby boy was kidnapped from a Lebanon, Kentucky hospital. The Kentucky State Police confirmed on December 12, 1985, that a baby, identified as the boy kidnapped from a bassinet in his mother's hospital room by a woman dressed as a nurse, was found in Hardin County, Kentucky. The person responsible for the kidnapping was seen by 12 hospital employees, but none stopped her because she was "disguised very well" and they thought she was a new nurse. Positive identification was made through the baby's footprints which had been taken shortly after birth.

Emergency rooms. In this area the security officer or hospital staff is likely to encounter belligerent patients and belligerent visitors. Arguments between family members and other parties to a dispute often continue into the emergency room. Often, people entering the emergency room are intoxicated, emotionally upset, or severely incapacitated. These people cannot be allowed to roam around unsupervised. Facilities such as vending machines, washrooms, and telephones should be available in the waiting room. The corridors and other areas of the hospital should be separated from the emergency area so that they can be controlled with a minimum number of supervisory personnel. Sometimes it is advisable to station a security officer in the emergency room. Furthermore, it is in the emergency room that a security officer is most likely to interact with law enforcement personnel. Provision should be made for the law enforcement officers to have a place to conduct interviews and communicate, at least by telephone.

Visitors and visitor control. There are several kinds of visitors to whom the hospital owes varying degrees of care. While these visitors may fall into categories of invitees, licensees or trespassers, they can also be placed in categories as visitors to patients, to employees, to administrative personnel, to purchasers of supplies and to other hospital functionaries.

There is firm consensus that patients need some visitors for their well being. With this in mind, visitors should be controlled so that the number and kind of visitors are balanced with the treatment and care required for the patient. Visitors should not be allowed to become a nuisance.

There are six areas of special consideration for control of visitors:

1) Psychiatric units.
2) Surgical units.
3) Intensive care units.
4) Pediatric units.
5) Obstetrical units.
6) Isolation units.

The most favored visiting schedule is twice daily with a two-hour limit for each period and two visitors at a time per patient. The times of visits should be in the afternoon and evening.

There is general acceptance of card, pass, or badge control for visitors. If any of these systems is used, it must be assured that the authorization to visit, be it pass, badge or card, not be lost or duplicated. It stands to reason that if the visitor is expected to wear a badge on the outside of his clothing, all personnel of the hospital must wear a comparable kind of identification. If they do not, all a visitor has to do to give himself the appearance of a hospital employee is to remove his badge. One innovative system is the use of a 5″ x 6″ card. It is difficult for the visitor to hide it or lose it. It should be laminated with the patient's name put on it with masking tape. There are many additions one can make to this system. For example, a different colored card can be used for different days of the week. Cards can be imprinted with instructions indicating that the card must be carried visibly in the hand and must be returned when the visitor leaves the hospital. Visitor rules and regulations can be printed on the back of the card also.

Nurses and candy stripers should be utilized to control the movement and circulation of visitors as much as possible. It is not recommended that security be assigned this task as a primary mission. Above all, it should not be the security officer who announces that visiting hours are over. Unit personnel should assume that responsibility. During nonoperational hours, the policy must be clearly stated as to how visitors may be allowed to enter the premises and who is allowed to approve the visits. All people who enter during those hours should be challenged. The perimeter doors must be secured by locks, alarms, or patrols. The night entrance should be at the emergency room. Portable desks or a second set of doors locked and manned by a security guard can be effective in controlling people who enter. This kind of sally port is called a mantrap in hospital security lingo. If the hospital has an attached physicians' office building, the connecting corridor may be used as an effective control point. There will always be exceptional cases and the security officer must be kept informed as to who might be entering. It might be that a relative or spouse of a patient will be allowed to visit any time; a visitor from out-of-town may be an exception; a change in the condition of the patient might be a cause to allow visits at unusual hours; or a person who needs medical care might be seeking admittance.

The relationship between patient and hospital is contractual. The hospital assumes a special duty toward the patient and his well-being. Negligence of any of its agents or employees can involve the hospital in embarrassing legal action and impose serious liability upon it. If a third person suffers injury or damage to his property because of the negligence of an employee who is acting within the scope of his employment, the hospital can become liable under the doctrine of *respondeat superior*.

A peculiar arrangement occurs in the hospitals in some states where attending physicians and private nurses are considered to be independent contractors and are not employees or agents of the hospital. In these same states, residents and interns are likely to be considered as falling within the doctrine of *respondeat superior*. Nevertheless, the hospital as a corporation may be negligent and subject to suits for damages if it maintains its buildings and grounds negligently, furnishes defective supplies, employs incompetent help, or fails to meet accepted standards of safety. Security will be in the forefront when lawsuits are mentioned.

Private hospitals do not employ totally proprietary security forces like those of major industrial corporations; instead, many have relied upon a contract security agency to supplement their proprietary force. Despite the increased attention given to security, we still find many situations where security is prevented, intentionally or unintentionally, from contributing its full potential to the health care field. This may be attributable to institutional leadership, security directors, or both.

Nursing Homes

Nursing homes generally are health care units for the elderly. A limited number of nursing-type facilities exist in major metropolitan areas to allow a patient to have restricted medical services, but not the full-service plan provided by most major hospitals. Cost and the need for the restricted service would be determining factors in this type facility. Similar to the hospital, the nursing home is faced with the security problems of visitor control and internal theft. Robberies are very prevalent, with the elderly occupant of the nursing home most often the victim. Most Americans are very protective of their elderly relatives and any criminal act where an occupant is the victim must be taken seriously by the administrators of a nursing home. Restricted access doors and CCTV are widely employed by nursing homes. Most nursing homes have security services provided by a contract security firm.

Retail

The retail industry is a part of society that the average citizen will have direct contact with on a regular basis. As in any commercial area, the more persons who come into contact with a particular facility, the more security problems that facility will experience. Loss of merchandise is the greatest security problem associated with the retail industry. In most cases, how and when the merchandise was taken is unknown. Retailing in most areas is a very competitive business and a few losses of valuable merchandise may make the difference between remaining in business and closing.

Shopping Centers and Malls

Shopping centers and malls are constructed to provide numerous retail stores in one large complex. Generally, at least one large department store will dominate with several small speciality shops in close proximity. Parking garages or open parking lots allow thousands of automobiles to park near the facilities. This congregation of people and automobiles in a relatively small geographical area creates a security problem. Customers are not going to shop at stores where thefts and robberies occur on a regular basis, leading to a further loss of revenue. The flow of pedestrian traffic is monitored in many shopping centers and malls by CCTV. This is not done so much as a security measure but to prevent any obstruction to the free access to the center by a customer. Shoplifting and thefts from a particular store are usually handled by the particular store. Cooperation is closely maintained between the security personnel of the shopping center and the security personnel of a store. Also, close contact is maintained with local law enforcement agencies.

After closing hours, fire watch and the prevention of burglaries are the primary functions of security personnel of a shopping center. This security service is generally provided by a contract security firm.

Retail Establishments

The loss of merchandise by retail stores is one of three major problems facing retail owners. (The other two are a lack of expertise in the particular field and a lack of capital.) The crime loss experienced by retail establishments is almost twice that of other industries such as manufacturing, wholesaling, services, and transportation. Shoplifting is the principal security problem associated with most retail establishments. Modern merchandising techniques create a significant security problem by emphasizing customer accessibility to merchandise. Even though major improvements have been made in electronic detection systems, shoplifting has continued to plague most retail establishments. Internal thefts by employees account for a significant percentage of losses to most retail establishments. In many cases, losses due to shoplifting and employee thefts are equal to or greater than the net profit of the business.

Most retail establishments operating on a regional and national level employ a proprietary security force. These employees have major responsibilities for planning, training and implementing all phases of various security operations. Since security services are closely related to safety services, many companies assign their security forces responsibilities in all areas of loss prevention.

Industrial

Some industrial companies must provide for the security of thousands of acres of land and a variety of seemingly unrelated activities. On the other hand, some industrial operations are very small, with just a few employees engaged in one basic activity. Regardless of the size of the operation, however, industrial facilities have common security problems related to the protection of their property, personnel, and information. In fact, some of the most progressive security practices are to be found in the area of industrial security, which is rapidly encompassing such activities as fire protection, traffic control, investigations, and the entire function of protecting life and property within an industrial enterprise.

Manufacturing

Manufacturing facilities are as varied in size as they are in the products they make. Security for a small, localized manufacturing plant with just a few employees may be relatively simple, while security for a huge multinational corporation may encompass many potential contingencies which often have implications that affect the entire United States. Security in the manufacturing industry must be concerned with theft of both raw materials and finished products, with the physical security of installations, with personnel security, and with the security of classified information.

The traditional "night watchman," while still apparent in many manufacturing facilities, is rapidly being replaced by security departments responsible for a wide range of loss prevention activities. This has encouraged the development of the security manager position whose duties include security, law enforcement, fire prevention and protection, accident investigation, public relations and other areas of security and safety.

Public and Private Utilities

The energy problem in America during the last few years has focused attention on the utility companies. In several foreign countries, terrorist organizations have been able to immobilize major parts of cities, and the same potential exists in this country. Sabotage of an electric power plant or a major substation may cause a blackout which could last for hours or days. A damaged natural gas facility or a water facility could cause problems for a large segment of a city, especially during severe winter months.

Nuclear facilities have been brought to the forefront of public attention with the nuclear accident at Three Mile Island, Pennsylvania. Not only are potential accidents a major security problem, but the attention given to all nuclear plants is creating a security problem that did not exist a few years ago. Environmental groups and ordinary citizens are demonstrating in large numbers against nuclear facilities in operation and under construction.

Most public and private utilities have a proprietary security staff which is supplemented by employees of a contract security firm.

Energy

Energy in America for years has been synonymous with oil companies. The oil and gasoline crises of 1974 and 1979 focused national attention on oil companies, and as a result, the Federal Department of Energy is requiring more security for manufacturing and storage areas. Energy facilities tend to be regionalized and the resultant need for security personnel likewise is regionalized. Coal is the major fossil fuel in America and more attention will likely be devoted to this resource in the future. Like oil reserves, coal is located only in a few states and will thus become a regional security problem.

Most oil companies are multi-national corporations, and so are subject to worldwide security problems. Kidnapping of executives has been occurring for several years in foreign countries and is potentially a major security problem today for American firms.

Special Events

In most major cities, private security plays a significant role in order maintenance and traffic control at special events such as football games, fairs and amusement centers. Local law enforcement agencies generally do not have the budget nor the personnel required to service such special events. On a national basis, a few contract security firms specialize in providing security services for all types of special events.

Civic Centers

Civic centers are multi-purpose facilities that are usually owned or controlled by local government. Their security problems are primarily order maintenance and traffic control. The major difference between one large crowd and another is the emotional excitement caused by the entertainment. Rock concerts arouse a different type of emotion than would be expected from a professional sporting event. A contract security firm generally has a contract with the civic center and provides security for all events.

Amusement Centers

The traveling, seasonal carnival is almost a thing of the past. It has been replaced, or at least seemingly so, by huge amusement parks located throughout the United States. These facilities cover hundreds of acres and include many separate structures. Depending on their geographical location, some of these facilities are open throughout the year.

Order maintenance, parking, and safety are the primary concerns of amusement center security. The need for a safe and orderly flow of thousands of visitors per day necessitates a security force designed to react to any contingency affecting the security of visitors or employees. Almost all amusement centers utilize a proprietary security force.

Fairs and Exhibits

Public facilities and properties for fairs and exhibits are often used throughout the year. The quantity and quality of security needed for each event is dependent upon the type of function and the number of people attending.

At most events, security problems would be crowd control, parking, and order maintenance. However, at certain events such as art exhibits or antique shows, physical security and the prevention of thefts would be of primary concern. During idle times vandalism and theft are problems, which in some facilities would necessitate a full-time security force. Many large fair and exhibition centers utilize a proprietary security force throughout the year, while others employ a contract agency when an activity requires such coverage.

Government

The security of government buildings and related real and personal property is an immense task. It requires that protection be provided for thousands of acres of land and buildings located throughout the United States. The vast number of government buildings and holdings makes it impossible for regular law enforcement units to provide service sufficient to satisfy all the demands for security. Some governmental agencies such as the Tennessee Valley Authority and the General Services Administration maintain large proprietary security forces. On the other hand, smaller governmental units at the state and local levels utilize either small proprietary security staffs or employ a contract agency.

Security problems for government lands and buildings range from vandalism and theft to the newer threat of terrorism. Security has arrived as a full and equal partner with the other services that are required to operate government buildings and facilities.

Tennessee Valley Authoriy

The Tennessee Valley Authority (TVA), headquartered in Knoxville, Tennessee, is the largest government-owned utility conglomerate in the United States. TVA operates all types of electric utility plants—hydroelectric, coal, and nuclear. In addition, TVA operates picnic areas and recreational facilities on most of its property. A huge recreational area is operated at the Land Between The Lakes facility in western Kentucky where there are provisions for hunting, fishing, camping, and boating. TVA has a proprietary security force which operates like a regular police force. TVA property is federal property and as a result most of the citations and arrests are handled in Federal Court.

Government Buildings

The General Services Administration (GSA) is charged with managing, operating, maintaining, and protecting federal buildings and related real and personal property. General Services Administration operates the Federal Protection Service Division and has approximately 5,000 personnel. The regular force of Federal Protective Officers is augmented by private guards on contract to GSA at locations and buildings which need more protection. Location will also dictate whether GSA utilizes the services of a private contracting agency or the regular GSA force. Many federal buildings are located in isolated areas which makes it impractical and uneconomical to utilize federal employees.

Federal Protective Officers are required to attend a service training academy and regular in-service training courses. Federal Protective Officers have the same powers of arrest as a regular police officer while on federal property under GSA control. The contract guards have only the powers and authority of private citizens.

Most states have similar security forces which have jurisdiction over security of state-owned, leased, and occupied buildings and property. The authority of such personnel is generally limited to the property being protected.

Public Housing

In the large metropolitan areas, housing authorities were established years ago to provide housing for low-income persons. With the advent of these multi-unit apartments in relatively small geographical areas, security quickly became a problem. Housing authorities are generally governmental or quasi-governmental and have a proprietary security force. The power of arrest is the same as the regular police force for the jurisdiction. Private housing projects have been constructed under the auspices of the Department of Housing and Urban Development and contract service firms handle the security services. Protection of life and property are the primary security problems associated with public housing projects. Thefts, assaults, and vandalism are the specific crimes that security officers must handle on a regular basis.

Schools

Colleges and universities have long recognized the need for their own security force. Most have a well-organized force comprised of high-quality personnel, and quality and quantity of officers have increased during the last decade. However, elementary and secondary schools outside of large cities did not recognize the need for security until recently. During the last few years, elementary and secondary school security programs have increased in number and size.

Elementary and Secondary Schools

On a nationwide basis, elementary and secondary schools are experiencing seemingly uncontrollable security problems. Vandalism has become so extensive that administrators have assigned teachers regular duties to patrol hallways and restrooms. Recognizing that teachers were employed to teach and not to handle security, especially discipline matters occurring outside the classroom, most school administrators have accepted a proprietary type security department. Even though safety of individuals is the primary concern of school safety officers, order, maintenance and vandalism are the problems which occupy most of their time. School security administrators have known for some time that security problems are reduced when a good discipline system is allowed to be maintained by the classroom teacher and the school principal.

College Security

Historically, a campus security program was attached to either the Dean of Students or the Director of the Physical Plant. If a college or university sustained troublesome or serious damage (more than violations of the student code of conduct), a night watchman or security officer was employed by the Director of the Physical Plant. Fire watch was the primary duty of the security guard or watchman.

In other cases, the deciding factor on whether to employ a security officer was the cost of the unlawful act. If the cost of replacing damaged locks, doors, windows, etc. reached a figure equal to or greater than the cost of employing a security officer, a decision would be made to employ the security officer. Up to this point, the college or university relied on the services of the local city, county or state police agency. If a college or university experienced law breaking by students on campus, security officers would be employed by and attached to the Dean of Students. In most cases, the decision to employ a security officer was made based on the number of students committing violations during off-hours and whether these violations resulted in damage to university property. Security was always a secondary function of the unit to which it was attached, whether the office of the Dean of Students or Director of the Physical Plant.

The development of the campus security officer has been largely determined by external factors. The appearance of the automobile and the campus building boom of the 1960's shifted the security officer from a fire watcher to a protective and control function.

Security under either of the above arrangements had serious problems: (1) the job and duties were seen as menial, (2) the benefits, pay fringes, equipment, training, etc. reflected this attitude and as a result were very low, (3) legal authority was uncertain due to the lack of statutory provision for a watchman-type position, and (4) there was a lack of job specifications as defined by the campus office responsible for security.

Since the shift in the function of campus security in the early 1960's and the problems of student unrest in the late 1960's, campus security has moved toward professionalism and placed more emphasis on personnel and training. State statutes that provide authority for campus security also require basic training and set minimum standards for employment. These standards are usually parallel with the statutes pertaining to police in that particular state.

It has been recognized that the quality of security personnel will affect the quality of their services. One characteristic of personnel which has undergone major change is the age of the individual. The stereotype of the campus watchman of the past was an old retiree who could barely walk, let alone keep the campus secure. Today, most campus security offices hire applicants who are between twenty-one and thirty-five years of age.

Educational requirements have also undergone change. In the past, a high school diploma (and in some cases not even that) was all that was required. At many institutions today, college work is a requirement for employment.

One of the major improvements that has been made in campus security during the last decade has been in training. Even though the type and amount of training varies from state to state, the common practice today is to follow the minimum standards set for police officers in that state.

Increased standards in age, education and training have brought an increase in salaries and fringe benefits. In most states, the campus security officer is compensated at a rate equal to or above the mean rate for other police officers. Those officers employed at state-supported institutions generally participate in a state-funded pension program.

Security Today

Confronted with alarming rises in crime and the constant fear of crime, Americans are reaching out for protection beyond that which can be provided by the nation's overcommitted and understaffed public law enforcement agencies. As a result, the number of private security personnel now exceeds that of public law enforcement, and the number of companies doing business in security products and services has increased dramatically, along with the number of businesses employing proprietary security forces. (figure 2-3)

This significant growth has resulted from the recognition of private security's potential for contribution to national crime prevention and reduction. It is here that the distinction between the roles of private security and public law enforcement is clearest. Public law enforcement forces are apprehension-oriented, that is, they usually act after the crime has already occurred. On the other hand, private security is prevention-oriented and emphasizes the prevention of crime. While their orientations are different, the goal is the same: the elimination of crime. Serving a common goal emphasizes the common interests of private security and the public law enforcement system. The private security sector cannot and was never intended to replace public law enforcement; instead, each complements the other in the effort to control crime. The scope and range of the nation's crime problem dictate that the two components work closely together without competition or disharmony. The current trend in private security toward increased professionalism will serve to upgrade existing practices and procedures. Higher standards of employment, training, and licensing requirements will increase the effectiveness of private security services and provide for a more mutually productive relationship between private security and public law enforcement.

Figure 2-3
Summary of Protective Service Worker Employment.

PUBLIC LAW ENFORCEMENT		**580,428**	**Percentage[7]**	
(Sworn Police Personnel)				
Local	495,842		85.4	
State	55,042		09.5	
Federal	29,544		05.1	
			100.0	
PRIVATE SECURITY		**678,160**		
Government Guard	**35,982**			**5.3**
Local	16,040		44.6	
State	9,130		25.4	
Federal	10,812[1]		30.0	
			100.0	
Contract Guards/Workers	**341,102[2]**			**50.3**
Proprietary Guards/Workers	**301,076**			**44.4**
Industrial/Manufacturing	65,800		21.9	
Construction	3,010		01.0	
Retail	55,838[3]		18.5	
Financial Institutions	16,874		05.6	
Real Estate	38,179		12.7	
Health Care Facilities	29,003		09.6	
Educational Institutions	25,553[4]		08.5	
Utilities/Communications	2,594		00.9	
Distribution/Warehousing	4,620[5]		01.5	
Hotel/Motel/Resort	8,773		02.9	
Transportation	5,109[6]		01.7	
Other	45,723		15.2	
			100.0	**100.0**
GRAND TOTAL		**1,258,588**		

Source: *National Industry-Occupation Matrix* 1980-90 Bureau of Labor Statistics, 1982.

1. includes 1825 "other" protective service workers
2. excludes 28,695 guards and alarm company armed response workers
3. includes 28,695 guards, 18,279 store detectives, 8,864 fitting room checkers
4. excludes organized police departments
5. includes wholesale trade and truck/warehousing (transportation)
6. excludes railroad police and transit police
7. modified table from: *Hallcrest Report: Private Security and Police in America,* William C. Cunningham and Todd H. Taylor, Chancellor Press

Discussion Questions

1. Outline the major differences between contract guard services and proprietary guard services.
2. Discuss the three basic types of alarm systems.
3. Discuss at least three of the specialized areas of security.
4. List three aspects of the Bank Protection Act of 1968.
5. Discuss why retail losses and thefts may have more significance to the average person than losses or thefts in other areas.
6. Trace the evolution of college security from its beginnings until the present time.

Chapter 3
The Threat Environment

To appreciate the task and role of security in America one must understand the tasks it faces. Any threat to security can be classified into one of two broad categories:

1) Those which occur naturally in the environment.
2) Those which are man-made, whether accidental or intentional actions.

Some of the hazards to security, such as fire and flood, may be natural or man-made depending upon the source or cause of initiation.

Natural Hazards

The evaluation of risks relating to natural hazards must be made from an analysis of geographical and meteorological records that pertain to a given location. Some natural occurrences are, of course, more common to one area of the United States than to another. The following list is representative of the more common natural hazards, and where they are most likely to occur:

1) Earthquake—Most common to the Pacific coast.
2) Tidal Wave, Tsunami—Coastal areas.
3) Flood—Coastal areas and low-lying areas having natural or man-made waterways.
4) Fire, Lightning—Anywhere in U.S.
5) Storm—Hurricane (coastal areas), Tornado (anywhere in U.S.), Snow and Ice (anywhere in U.S. except southern areas), High Winds (anywhere in U.S.), and Temperature Extremes (anywhere in U.S.).

Modern technology has made most of these natural hazards somewhat predictable. Often it is possible to have several days' warning and some indication of the probable magnitude of the pending hazard. However, natural hazards can also strike without warning, and it is at such times that pre-planning and immediate, adequate reaction capability are of the utmost importance.

While it is generally impossible to prevent such natural disasters, there are steps that can be taken. First, choose a site for the facility where such disasters are rare. Second, construct a facility that will withstand localized natural hazards. Third, develop emergency plans to reduce the damage that such disasters can do.

Man-Made Hazards

Man-made hazards may be either intentional or accidental actions. Often it is difficult to determine if an occurrence was intentional, an accident, or even if it was the result of a natural phenomenon. A flood, for example, may be the result of too much rain in a very short period of time, or the result of an intentional or accidental act in the improper operation of flood control barriers.

While it is beyond the scope of this text to address all loss-producing occurrences, it is important that crime and fire, the two most common and destructive threats, be discussed in detail.

Fire

Throughout the United States and the world, fire takes an enormous annual toll of human life and property. During recent years, fires have caused an annual toll of some 12,000 deaths in the United States alone. In addition, thousands have been injured, millions of dollars in property have gone up in smoke, and untold indirect losses of jobs, customers, employees and businesses have occurred. Why is fire such a destructive force in a society having such enormous technological and engineering capability as ours? The answer is simple, according to the National Fire Protection Association. Fire is more related to human acts of omission and commission than to science. There is no denying that people cause most fires, and that most "people-caused" fires are due to thoughtless acts of carelessness than from any uncontrolled interactions of reactive materials.

The threat of fire, then, is generated by human acts such as poor housekeeping, careless use of smoking materials, inadequate or improper construction, utilization and maintenance of equipment, and intentional acts of arson and sabotage.

Every environment has certain hazardous qualities, with some being more fire resistant than others. Fire prevention programs must include an analysis of the hazardous materials and operations that are common to the environment. This analysis must then be used to develop a set of workable fire safety regulations.

Fires caused by human error can occur in a variety of ways, most of which are preventable. Training and education are the keys to fire prevention. All employees of an organization should be fully informed of the importance and techniques of fire safety. Company rules and regulations regarding fire safety should be common knowledge to every employee. These standards should be constantly re-evaluated and enforced to the fullest extent.

It would be naive to think that the threat of fire can be totally eliminated. The threat of fire remains, regardless of the pre-planning, education and training that is accomplished. The goal of total prevention of fire, while not obtainable, must always be sought. The burned building, the charred inventory, and the dead or injured are only the obvious and direct losses of fire. Not so visible are a variety of indirect losses: (1) temporary and often permanent loss of jobs; (2) loss of company sales through inability to provide products; (3) permanent loss of some customers who switch to other suppliers and never return; (4) mental anguish suffered by friends and family members of fire victims; (5) inability to recover from fire losses and subsequent bankruptcy; (6) loss of confidence in the company's ability to provide a safe work environment; (7) lawsuits involving company negligence and subsequent unfavorable judgments; and (8) increased insurance rates.

Fire, then, is a real and major adversary, capable of causing enormous losses and hardships. The function of security in both its prevention and control is direct and critical. Training, education, evaluation, and inspection are areas in which security can play a vital role.

Crime

Crime is behavior that violates the criminal law. Criminals are those citizens in our society who have engaged in acts that have been socially and officially defined as exceeding the limits prescribed for legally acceptable behavior. Virtually all modern societies have enacted criminal laws to cover certain basic violations, such as homicide, injurious assault, incest and violations of property rights. However, there is often considerable variation as to the exact definition of offenses, circumstances accepted as extenuation, severity of official reaction and societal sanction. The United States alone has fifty-one (state and federal) legal jurisdictions, each with its own set of criminal statutes.

Crime is one of the principal threats facing all Americans in the United States. An enormous variety of acts make up the crime problem. Crime is not just the street-tough teenager snatching an old lady's purse. It is a professional thief stealing cars "on order." It is a well-heeled loan shark taking over a perviously legitimate business for organized crime. It is a polite young son who suddenly and inexplicably murders his family. It is a corporate executive conspiring with competitors to keep prices high. It is a bright young socialite who, for a perceived cause, can participate in violent acts of terrorism and destruction. No single theory, formula, or definitive generalization can explain the vast range and scope of human behavior called crime. Its causation is multiple and complex, and to even begin to understand its intensity and scope one must study and evaluate the threat environment, the criminals and their functional roles within a given frame of reference.

Individual or group actions that can be classified as crimes are divided into two basic groups: (1) crimes against persons, and (2) crimes against property. Crimes against persons include murder, kidnapping, rape, assault, robbery, etc. Crimes against property include burglary, arson, vandalism, sabotage, shoplifting, auto theft, etc. Crimes are also classified as misdemeanors or felonies. A misdemeanor is a lesser act usually defined as a crime with a penalty of a fine and/or up to a year in jail, while the penalty for a felony is a fine, one or more years in prison, or the death penalty.

The major divisions of the law are *civil* and *criminal.* Civil law deals with legal liabilities between individuals, while criminal law deals with crimes that the government prosecutes in its own name. Private security personnel, while not expected to be lawyers, must be familiar with the law, civil and criminal, as it relates to the performance of duties. The manner in which private security personnel perform their duties has a direct bearing on the attainment of industry goals and professional status.

Criminal acts must be understood from the standpoint of corresponding legal definitions. The following are generalized definitions of some of the more common criminal acts which are of primary concern to security personnel. However, be aware that crime definitions may vary somewhat from state to state, particularly as to degree and penalty.

Robbery – The stealing or taking of something of value from the custody or control of a person by force, threat of force, or by putting that person in fear of his/her welfare.

Burglary – The unlawful entry of a building or structure to commit a theft even though no force may have been used to gain entry.

Larceny – The unlawful taking or stealing of property or something of value without the use of threat, violence or fraud.

Arson – The willful, malicious burning of a dwelling, building, or other property with or without the intent to defraud.

Fraud – Intentional misrepresentation to induce another to give up or part with something of value or to surrender a legal right.

Vandalism – The intentional or malicious destruction, injury, or disfigurement of property.

Shoplifting – The removal of merchandise from a store with the intent to deprive the owner of his property without paying the purchase price.

Homicide – The killing of one human being by another human being.

Assault – An attempt or threat, with force or violence, to do corporal harm to another.

Criminal Trespass - Disturbance of possession of the premises of another.

Embezzlement - Fraudulent appropriation of property.

The means by which these crimes and others can be perpetrated are varied and complex, but one can reduce the risks of being victimized by being aware of some of the factors involved. For example, one should be aware of how types of business, places of business, types of items carried, nearness and responsiveness of police protection, and types of protective hardware and procedures employed are determining factors in inviting or deterring criminal attacks.

No one can be certain, regardless of the steps taken, that they will be totally safe from crime. After all, there will always be some deviant behavior in every society. Still, the man-made hazards classified as crime can be reduced to an acceptable level if appropriate measures are taken to deter, detect, and/or deny those activities that are detrimental to all. Later chapters will deal with specific techniques of deterring and preventing certain crimes.

The Criminal

Individuals identified as criminals, i.e., those that have been found guilty of a crime, are not readily identifiable in appearance, speech, manner, background, attitude, behavior, skills, or method of criminal operation. However, there are some identifiable factors that can be utilized to deter or thwart a criminal's attempts to commit a crime, regardless of his personal or physical characteristics.

In the past, various criminological theories have been proposed to explain why people commit crimes. These explanations have been physiological, sociological, psychological, and even religious and political. However, current thought seems to accept the idea that an individual is a product of his own psychological make-up and the environment to which he is exposed. Thus, there is no typical criminal, because each person acts and reacts according to his own inherent and acquired characteristics and capabilities.

The primary objective of most criminals is personal or financial gain; thus, currency, objects convertible to cash, and objects taken for personal use are most often stolen. Other criminal objectives, though not as common as the above, are the destruction of property, social or political change, revenge, satisfying mental or psychological compulsions, etc. The property owner, then, must evaluate what he has in terms of how he can best protect it against the criminal who may desire what the owner has. For example, cash or property easily converted to cash is more desirable to the criminal than something which he cannot sell or cannot use.

It is in this way that homeowners, businessmen, and others must take steps to remove or decrease the opportunity for criminal activity. Most criminals do not want to get caught. Targets of opportunity for robbery, burglary and other crimes exist where it is readily apparent to the criminal that his objectives are easily accessible and relatively unprotected and that there is little danger of being detected or apprehended.

The extremes of criminal proficiency and expertise are evidenced by the amateur and professional. The amateur does not depend on crime for his livelihood. Instead, most of his illegal deeds are committed on the spur of the moment as the opportunity presents itself. While no one knows for sure, it is assumed that professional criminals account for only a small percentage of the total criminal population. If this is true, then most crimes are committed by amateur criminals and most crimes are crimes of opportunity. Put another way, it is the inexperienced, nonprofessional "crook" that commits the majority of criminal acts. In fact, current crime statistics show that more than fifty percent of those arrested for criminal acts are juveniles.

The professional criminal, on the other hand, is one who makes his living through crime, usually one particular type of crime which he considers his specialty. Of course, the professional began his career as an amateur, but was able, through luck or perhaps intelligence, to develop his skills to the point of being a successful criminal.

Given the time and proper conditions, a highly skilled and determined professional criminal can successfully penetrate nearly any protective system no matter how complex or strongly fortified. In general, however, if a residential or commercial facility is secured by strong locks or hardware and other perimeter barriers, is well lighted, and is protected by an appropriate type of alarm system, most criminals, amateurs or professionals, will seek other, easier targets.

It must be recognized that many criminal activities are initiated from within the environment being protected. The threat, represented by the employee, means that the owner/manager must evaluate his protection needs from both within and without. Employee thieves range from those who steal a little money or small items every now and then, to those who systematically steal and bleed the company of huge amounts of money or inventory. Some will steal only when inadequate security seems almost to invite it, while others will manipulate and seek ways to steal regardless of what controls are in effect. The majority of dishonest employees will use very straightforward, basic techniques of stealing while a few will utilize very sophisticated methods of draining off company assets. In any case, it is quite apparent that it is the employee who has the greatest exposure to company assets and the opportunity to engage in numerous types of behavior detrimental to the company.

The motivating factors of each type of honest or dishonest employee exist to varying degrees. The reason or combination of reasons that motivates one employee to commit a criminal act may or may not motivate another employee to do the same. While no company wants to think of its own employees as being dishonest, it must take steps to deal with employee crime as a condition of doing business. In order to protect his property, the owner must in effect view his belongings as though he himself were a criminal, employee or nonemployee, looking for an easy "hit." The view from the criminal's persepective should reveal both the weaknesses and the strengths of existing security measures. Effective countermeasures can only be taken when the situation is understood. It is not an easy task but is one that must be taken if goals and objectives are to be attained. Current information suggests that losses each year to employee crimes are increasing, and are much larger than those due to robbery, burglary and shoplifters.

Computer Crime

One type of crime, computer crime, cannot be approached via the traditional methods of detecting, investigating, and preventing losses. The rapid technological developments of the computer age have created a high-tech environment for a new kind of crime, i.e., the computer crime. Within the span of a few years, the computer has become a commonplace yet vital fixture of business and industry. Perhaps no single factor in business or industry can surpass the computer in its potential to harm or destroy an entire organization. The compromised computer, whether by direct or remote manipulation, by theft of storage tapes or discs, by sabotage or accident, affects the ability of the user and owner to operate efficiently and effectively. The confidentiality and integrity of information and data stored in the computer is, in many cases, absolutely essential to continued, competitive operation. The theft, destruction, or manipulation of information and data compromises and sometimes destroys the ability and capacity to simply stay in business.

Terrorism

Terrorism is by no means a new form of warfare; it has been evident down through the centuries in many conflicts of ideas, wills, and national interest groups. Yet, prior to World War II the victims of terrorism were nearly always specific individuals singled out for death either in retribution for some action or because of the position of leadership or authority they held. In the past few years, however, terrorist acts have often been characterized by seeking out vulnerable and defenseless targets. The trend has been complete disregard for noncombatants such as women, children, and travelers. Terrorists have demonstrated that they can have a crippling impact on national and international events far beyond what their limited numbers and resources would suggest. Terrorism has been used to focus the world's attention on various causes, announcing the existence of certain perceived grievances or problems. Governments have been embarrassed, compelled to grant concessions and pay huge ransoms, and in some cases capitulated to terrorist forces. Governments and private interests have been forced to spend billions of dollars on elaborate security measures. In short, the revived and innovative crime of terrorism has reduced the quality of life in several parts of the world. There is little question that terrorism presents many unusual challenges; it is a powerful force from within and without that must be reckoned with and controlled.

Conclusion

Security, then, to be complete, must be approached systematically. Analysis of the threat, whether it be man-made or natural, is a mandatory step that must be taken before effective preventive or deterrent action can be accomplished.

Current developments in detection and alert systems have made it possible to have advance warnings of many probable or actual hazardous environmental conditions. With planning and adequate preparation, property losses and life-threatening events can be minimized.

The current crime situation indicates that we are confronted by a social problem of the most serious proportions, which threatens the welfare of all people, homes, and places of business in the United States. The heavy burden of preventing or deterring criminal acts, whether committed by the professional, the amateur, the employer or the non-employee, falls on everyone.

Discussion Questions

1. Briefly list the more common natural hazards and the locations where they are most likely to occur.
2. Identify and give examples of the two basic types of crime.
3. List several indirect losses associated with fire.
4. Describe the differences between the amateur and the professional criminal.
5. Define crime and identify its range and scope.
6. What is the purpose of criminal law as compared to civil law?
7. Why is it so important to provide protection and security for the computer?
8. What has been the effect of terrorism?

Chapter 4
Perimeter Security

The ability to protect and secure any facility or building largely depends upon the environment or general location of the structure. That is, the area immediately surrounding the facility must be secure if the facility itself is to remain secure. This area immediately adjacent to and surrounding a facility is the perimeter. Often, the value of the goods within a facility or the confidentiality of a business or industrial process therein requires that the perimeter be secured. This is accomplished through the installation of alarms, barriers, and lighting, the security of the structure itself, and the introduction of guards.

In many instances the security of a facility declines as accessibility to that facility increases. Also, as security precautions are increased, the cost to management increases, as does the inconvenience to employees, patrons and management. Therefore, the efforts to secure a facility perimeter largely depend upon that which is being secured. That is, there are tradeoffs in terms of amount of security with regard to costs, inconvenience, and aesthetic qualities. The ultimate selection and deployment of a security device is, therefore, dependent upon management's and security's perception of costs, needs and utility of action.

This chapter examines the problems and solutions to perimeter security. Issues include: site layout, perimeter barriers and protection, building surface security, cooperative issues, and perimeter security hardware and personnel needs.

Site Layout

Too often, security is only of minor concern during the planning and construction of a facility. Management is frequently more interested in other requirements such as safety, economy of construction and operation, and convenience. Only after the facility has been constructed and put into operation does management consider security. But by this time, the security problems are much harder to solve. Therefore, in order to enhance security, security specialists should have input during the initial architectural planning stages. This way, there can be a compromise between cost, convenience and security issues. If adequate attention is not given to security during the planning stages, then security costs or loss and shrinkage will probably increase.

One of the most critical factors with regard to site layout is the positioning of the facility on the building lot. Great care should be taken to ensure that the structure receives maximum exposure from adjacent thoroughfares. The more isolated from public view, the more likely that the structures will be susceptible to unauthorized entry or exit. A study by the Law Enforcement Assistance Administration showed that for 70 percent of the burglaries reported in selected areas of California, the point of entry was not visible.[1] Robberies also become a problem when there is limited visibility. Would-be robbers tend to select targets which are isolated or not easily viewed by pedestrians and vehicular traffic.

In most cases the facility should be constructed in the middle of the building lot so that movement around the facility is unimpeded and all four sides can be easily observed. When this is not possible and the structure must be located immediately adjacent to another structure, special security precautions must be taken to guard against surreptitious entry or exit.

Landscaping can also play an important role in reducing burglaries and inventory shrinkage. In most cases, management desires to make the facility as attractive as possible in order to enhance their business. This usually entails the introduction of shrubbery, trees and other vegetation or ornamentation. Although very appealing to the eye, landscaping makes it easier for security to be compromised. If possible, large quantities of vegetation should be avoided and should not be located within 50 feet of the structure. This will increase visibility and help deter illegal activities.

Special care should also be taken with regard to where merchandise, supplies and materials enter or leave the facility. These areas should remain free from obstruction to decrease the possibility of employees or carriers hiding goods and merchandise and later retrieving them. This is especially important around loading docks and rail entrances.

Finally, an integral part of landscaping is lighting, which is used to beautify as well as increase security. Frequently, management uses lighting for advertisement, without regard to security. (The problem of security lighting is addressed in a separate chapter).

The initial design of a structure will have a great deal of bearing on the security efforts. Buildings, facilities and perimeter areas must be not only pleasing to the eye, but also constructed so that they are not vulnerable to illegal activity.

Types of Physical Barriers

Physical barriers are of two primary types: natural and man-made. Natural barriers include bodies of water, mountains, cliffs, deserts, canyons, swamps, or other types of terrain which are difficult to travel through or over. These features of nature can be utilized to serve as primary or secondary barriers to unwanted or unauthorized entry. Man-made or structural barriers include fences, walls, grills, and bars.

Regardless of type, properly used barriers can effectively accomplish the following security objectives:

1) Define property boundaries.
2) Deter entry.
3) Delay and impede unauthorized entry.
4) Channel and restrict the flow of persons and vehicular traffic.
5) Provide for more efficient and effective utilization of security forces.

Perimeter Barriers and Protection

Perimeter protection is considered the first line of defense against unauthorized intrusions and the last line of defense against unauthorized exits. When constructed and operated properly, a perimeter barrier is a physical and psychological deterrent to unauthorized movement to and from the facility. A perimeter barrier deters thefts, intrusions and vandalism. It should also be remembered that the perimeter barrier will not stand alone as a total defense, but must be supplemented with guards, alarms, cameras, etc. Sometimes walls are constructed, but most often fences are used as perimeter barriers. Walls can be constructed out of wood, stone, cement blocks or concrete. The advantage of walls over fences is that they are stronger, and thus more resistant, to intrusion (especially stone or concrete). They can generally be constructed to any desirable height, and they can be constructed so as to add aesthetic qualities to the structure. The disadvantages to walls are that they restrict perimeter visibility (thus negatively affecting security), and they are much more costly to construct than conventional fencing. Walls are frequently used to secure smaller compounds where a high degree of security is needed and when a premium is placed on aesthetics.

Types of Fencing

Fencing is generally used to secure larger areas. There are three basic types of fencing which are traditionally used to secure areas: chain link, barbed wire and concertina wire. The type used depends somewhat on the permanence of the fence. Chain link fencing is usually used to secure permanent facilities, barbed wire is used for less permanent facilities or to mark perimeter boundaries, and concertina wire is used in emergency or short-term situations.

Chain Link Fencing

A chain link fence is attractive due to its clean, neat lines. Chain link fencing poses less of a safety hazard because it does not have barbs, yet the small openings still prevent intrusion. Finally, it is easily and inexpensively maintained.

Where a chain link fence is used, it should be constructed of No. 1 or heavier gauge wire with mesh openings no larger than two inches across. The fence should be at least eight feet in total height. The fence should be topped with a "V" top guard, or 45° shaped arm bars and three strands of barbed wire attached on both sides to inhibit unauthorized entries or exits (figure 4-1). A newer variation of the barbed wire top guard is to utilize

Figure 4-1
Chain Link Fence with "V" Shaped Topguard.

"razor ribbon," a thin ribbon of very sharp metal on which are attached razor-like projections of metal. The fence should be permanently attached to metal posts which are sunk in concrete. When possible, the fencing should be buried at least two inches (this is especially true if the soil is soft or sandy), or a strong wire should be woven through the lowest sections of the wire mesh to inhibit attempts to go under the fence. Where the fence crosses rugged terrain such as streams, hills, or ditches, precautions must be taken to ensure that there are no unprotected openings beneath the fence.

Barbed Wire Fencing

Barbed wire is seldom used to secure perimeters due to its unsightliness and the danger of inflicting wounds on those who come into contact with it. When barbed wire is used to mark boundaries, it should be five feet high and consist of three or four strands tightly stretched, attached to posts which are from six to ten feet apart.

On occasion barbed wire has been used to secure restricted areas. Great care should be taken to ensure that it is not vulnerable to compromise. Barbed wire fences

> . . . should be not less than seven feet in height plus a top guard, tightly stretched, and firmly affixed to posts not more than six feet apart. Distances between strands should not exceed six inches . . . the bottom strand should be at ground level to impede tunneling, and the distance between strands should be two inches at the bottom gradually increasing to six inches at the top.[2]

Barbed wire is not recommended for securing perimeters because of the aforestated problems. However, barbed wire on occasion is useful in supplementing natural barriers. For example, barbed wire could be strung along the side of a cliff to further deter intrusions.

Concertina Wire Fencing

Finally, concertina wire, barbed wire or the newer "razor ribbon" can be rolled into a coil and clipped together at intervals and used as a barrier to secure a perimeter. Concertina is most effective when one roll is laid on top of two other rolls, giving the barrier a height of approximately six feet. When it is utilized, the rolls should be fastened together and attached to the ground periodically with ground stakes. For the most part, concertina wire is one of the most difficult barriers to penetrate since it is extremely flexible and exists in large quantities.

Concertina wire is unsightly and hinders gound maintenance, and thus, it should not be used as a permanent barrier. However, concertina wire has its advantages and can play an important role in perimeter security. It can be laid and picked up rather easily by one person; thus it is an extremely mobile barrier which can be deployed for emergency situations. For example, if the permanent barrier, say a chain link fence, is damaged as a result of an automobile accident or a tree falling onto the fence, concertina wire can be easily laid to hinder intrusion until the permanent barrier can be repaired. Also, it can be used to temporarily block roads or paths in case of an emergency. Thus, concertina wire should be stocked for emergency situations.

A barrier should be constructed in as straight a line as possible to prevent intruders from hiding close to the fence. At least 50 feet should be cleared of structures and obstructions on either side of the barrier to prevent the hiding of persons, burglary tools or property. Such a clean zone provides for adequate observation on either side of the barrier. Finally, the barrier should be patrolled to check for obstructions and breaks in the barrier and to supplement the effects of the barrier itself. The frequency of patrol depends on the desired degree of security.

Building Surface Security

The second line of defense against intrusion is the building itself. A perimeter barrier may not always deter the determined intruder. This is especially true where perimeter barriers are located in isolated areas.

A building or structure should be considered an entity which can be intruded from six different directions: the roof, flooring, or one of four sides. Though efforts should be made to completely secure the exterior of the structure, the primary consideration here is roof access, doors and entrances, windows and miscellaneous openings such as fire escapes, vents, or delivery or trash portals.

One of the most vulnerable sections of any structure is the roof. The roof, especially flat roofs, are difficult to observe, thus giving intruders ample time to make an entry, and roofs are generally constructed of materials such as wood, tar and shingles which are easily compromised. If possible, roofs should be construced with a high pitch, which makes them difficult to maneuver on and easy to observe from the ground. Second, the number of attachments to the structure (such as fire escapes) should be reduced, making it more difficult to obtain access to the roof. Finally, the area immediately adjacent to a structure should be clear of obstructions so that it would be difficult to hide ladders, ropes, etc. If access is reduced, the probability of illegal entry is also reduced.

A large number of illegal entries and exits occur through windows. Windows are probably the most vulnerable part of any structure. If possible the number of windows should be reduced, and all windows on the ground level eliminated. Of course, this is not possible for certain businesses, and in these cases, efforts should be taken to reduce possible forced entry.

One common practice in business security is the introduction of metal grates or bars over windows. However, these precautions are frequently inadequate because of flaws in installation. When installed improperly, bars or grating can be pried easily from the face of the building. Proper installation includes steel connector bolts completely through the wall and connecting the bolts on either side with a piece of flat steel and nuts with lock washers (figure 4-2).

Figure 4-2
Protective Metal Grate on Ground Level Windows.

Many factories and businesses rely on steel-framed windows with steel strips holding each pane. If such windows are used, care should be taken to ensure that the lock cannot be maneuvered from outside through a broken pane of glass. Moreover, these windows only offer minimum resistance to a good prying device.

For the most part, protective windows and window coverings are helpful in deterring the nonprofessional criminal. This is an important step since the Federal Bureau of Investigation's Uniform Crime Reports consistently point out the large numbers of burglaries and larcenies that are committed by juveniles or nonprofessional criminals.

Doors are also frequently used to obtain illegal entry. Most often, wooden or unprotected doors are the primary targets. All doorways should be adequately lighted and free from obstructions. Metal doors are best, because they are most difficult to penetrate. When a wooden door is used, the inside should be covered with a steel sheet to prevent intruders from kicking or cutting a hole in the door. When possible, the door should be secured with a standard deadbolt lock and a horizontal retaining bar. Retaining bars should be installed on all doors except those most often used.

Finally, special precautions should be taken with door frames and hinges. Frequently, fire codes require that doors open outward, thus causing the hinges to be exposed to the outside and making the door vulnerable. In this case, special hinges should be used to thwart illegal entry (figure 4-3). If the door frame is constructed of wood or similar materials, it may be possible to pry the door open by springing the door loose from the frame near the locking mechanism. A common automobile jack can also be used to spring a door. Therefore, the door frame should be constructed of solid materials and possibly supplemented with an alarm device.

Figure 4-3
Pinning a Door to Prevent its Removal.

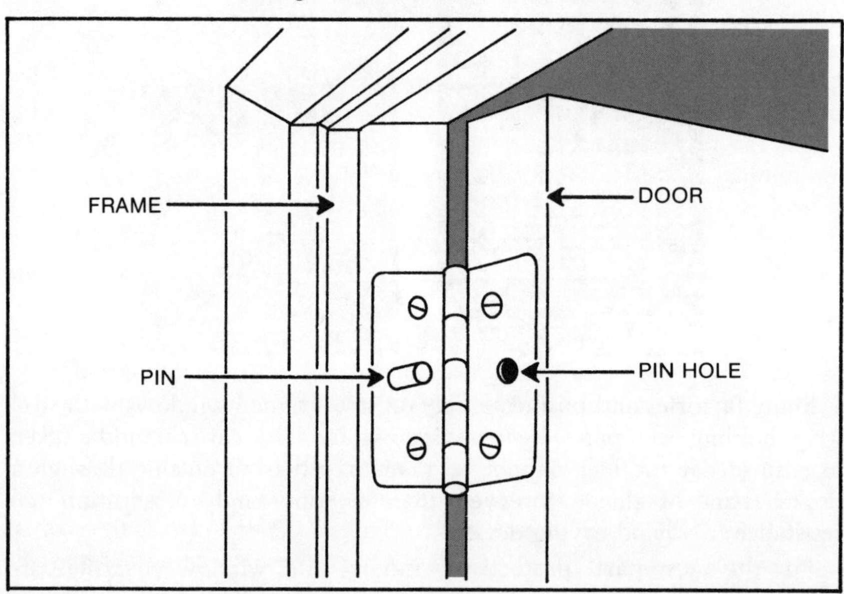

Garage or overhead doors not controlled by electric motors should also receive special consideration. These doors should be constructed of metal when possible. The lock should consist of solid sliding metal bars on either side of the door, secured with padlocks to inhibit unauthorized exits as well as entries.

With any structure there are always a number of miscellaneous openings such as fire escapes, vents, storm sewers or trash portals. These openings should also be secured. Even the smallest opening may jeopardize security for a building. For example, a common practice is to find or make a small entry point and allow a small juvenile to go into the facility to gain access for others. This is a reason why all exits, safety permitting, should utilize locks which require keys on the inside as well as the outside.

Finally, special precautions must be taken where buildings adjoin other structures. Whenever this occurs, it is possible to gain entry to a second building by knocking a hole through the common wall. This frequently occurs where a low security structure adjoins a high security structure containing valuables. The entry usually occurs during the weekend when the intruders have ample time to work unnoticed. If this situation exists, then security must either be supplemented with alarms, or particular attention should also be paid to the adjacent structure.

Perimeter Access Controls

Perimeter security measures, in addition to and in concert with perimeter barriers, must satisfy the needs and goals of the organization being protected. By its very nature, an organization is dependent on people for its success and existence. Any effort to restrict or control the movement of people through its gates and doors must complement the daily activities and operations of the facility. Perimeter security controls must allow for normal entries and exits and preclude unauthorized entries and exits.

Access Control

Effective and efficient access control, that is, controlling the coming and going of people and vehicular traffic, is a blend and balance of people, equipment, and procedures. The ways in which access control can be accomplished vary from simple to sophisticated, depending on the level of security required, from a lock and key or a security employee checking identification badges to magnetically encoded cards that operate electronic locking devices, closed circuit television to view individuals seeking access, or devices that utilize a computer to "recognize" an authorized individual by his handwriting, fingerprints, hand geometry, facial features, etc.

Thus, a corollary to the need for access control is the need for a system of identification to determine who is authorized for entry and who is not. Access lists, personal recognition, security identification cards and badges, badge exchange procedures, and automated electronic devices are among the methods used to identify and control the movement of employees, vendors, visitors, and clients.

The determination of what to use, where, when, and how to use a particular perimeter security device or procedure is dependent on the following considerations:

1) Existing perimeter barriers – fencing; walls; natural barriers; number, location, and type of perimeter openings.
2) Organizational type – manufacturing; retail; health care; transportation.
3) Organizational attributes – size; location; schedule of operation; goals.
4) Management – philosophy of management; managerial guidelines and policies; managerial support.
5) Employees – employee acceptance; employeee training; number of employees.
6) Risk factors – environmental variables; crime problems; vulnerability assessments; criticality factors.
7) Resources – money; equipment; manpower.

Obviously, there is no universal solution to perimeter security. For example, a hospital environment has different needs and objectives than a manufacturing facility. While there will be a similar application of basic security principles and concepts in both locations, the manner and measure of their utilization will be very different.

Parking

When a business or facility maintains a perimeter barrier and provides parking for clientele and employees, the parking area should be located outside the perimeter barrier. This greatly reduces the chances that intruders will enter the facility since it is easier for guards to monitor persons when they must pass a security point on foot. Such an arrangement provides excellent security and maximum control.

When management provides parking facilities, they should be immediately adjacent to the perimeter barrier and guard posts. The area should be well illuminated. This allows guards maximum observation over the facility and deters thefts and vandalism without having to deploy additional personnel in the parking area. Moreover, if mobile patrols are used at the facility, they should periodically patrol employee parking lots to enhance deterrence.

Vehicular Traffic Control

It is extremely important to not only control the comings and goings of employees, but also to control all vehicular traffic within the facility. The fewer entrances there are, the easier it is for security to control the perimeter. Thus, only those entrances necessary to the facility should be opened. In most cases, perimeter barriers have primary entrances, those used on a regular basis, and secondary entrances, those used for special occasions such as deliveries to isolated warehouses, maintenance, etc., or for emergencies such as fires or evacuations. In order to maximize security, guards should be posted at primary entrances unless they are secured or closed. Secondary entrances should remain secure or locked, and patrolled on a regular basis to guard against intrusion. It is seldom necessary to post a permanent guard at a secure secondary entrance.

Entrance or gate security is especially important in relation to shipping, receiving and disposal. When receiving shipments, gate security is particularly helpful in directing facility traffic. Additionally, if there is need for tighter security, an escort can be provided for incoming carriers at the gate until they reach their destination or leave the facility. Guards can also be used to monitor shipping by comparing shipment invoices with loads, either on a continual or random basis. This is especially important in a facility where expensive goods are produced. Such a procedure inhibits drivers and plant employees from falsifying invoices. For example, employees in a Kentucky firm were recently found to be using this method to steal truck axles. The axles were obtained through falsified invoices and sold to coal trucking firms.

Another problem which confronts security is the disposal of waste and trash. Frequently, employees or other persons hide merchandise, goods, etc., in trash bins where it is later retrieved. Therefore, periodic security checks should be made of trash and disposal areas, and vehicles carrying trash should be checked periodically as they leave the facility. Such precautions will reduce employee thefts and aid in maintaining control over the facility compound.

At all times security must work closely with management to marry the needs of management and the needs of security. It is important for the security staff to constantly monitor the facility's operations so that adjustments can be made in security as changes in the operations of the facility occur.

Perimeter Security: People and Hardware

When designing security systems, the security manager has a number of alternative protection devices and personnel at his disposal. He must thoroughly analyze the costs, benefits and problems associated with each alternative and choose those which best meet the needs of the specific facility. No one system is perfect for every job, and the manager must take particular care during this planning stage to analyze his needs.

Security Guard Personnel

In most cases perimeter barriers are supplemented with guard personnel. They serve to boost the deterrent effects of the perimeter barrier and to monitor the barrier for defects. The posting of sentries or stationary guards and the use of mobile patrols is largely dependent upon the nature of the facility being guarded and the degree of security desired. In cases where the perimeter barrier encloses an extremely large area, then it may not be efficient to post stationary guards to observe the total barrier structure. In isolated or semi-isolated areas protection may be reduced to infrequent mobile patrols or only an occasional maintenance check. However, if a high degree of security is desired it may be necessary to increase patrols and/or stationary posts, or it may be more cost-effective to construct a second or inner perimeter barrier closer to the area requiring the high level security.

Although guards are highly reliable as a means of enhancing security, they are extremely expensive compared to other security measures. Therefore, if possible, the guard force should be reduced to the lowest effective level via the use of other security devices such as perimeter/interior alarm systems, clear zones, high deterrent barriers, and security lighting. (See Chapter 5 for discussion of security lighting)

Closed Circuit Television (CCTV)

While not an intrusion detection system, closed circuit television is very useful in accomplishing physical security. Placement of television cameras at critical locations can provide direct visual monitoring from a centralized vantage point. Closed circuit television can be used to monitor gates, employee entrances, loading docks, receiving areas, and hazardous operational areas. A CCTV system ordinarily consists of a television camera, monitor, and electrical circuitry. Additional equipment might include a pan and tilt unit which gives remote control of the camera, an automatic scanning mechanism for the camera, or a video tape unit to record on tape what the camera sees. A CCTV system may be composed of one camera and one monitor or several of each. The sophistication of the system can be enhanced by utilizing various kinds of camera lenses, full color equipment, and/or by using cameras with the capability of producing clear images under minimal light conditions.

Since the primary means of providing perimeter protection is personal observation, CCTV offers a means of increasing the surveillance capability of security personnel. Visible CCTV camera units also serve as a psychological deterrent to would-be intruders. Their presence increases the risks for the potential intruder that any attempt to enter the facility would be observed.

Perimeter Intrusion Detection

Perimeter intrusion detection systems are used to provide electronic surveillance of established perimeter lines. Such systems can be employed to provide continuous surveillance of perimeter lines to signal the entry of persons into a protected area. The decision to use perimeter intrusion detection will depend upon various factors:

1) Characteristics of the environment to be protected.
2) Vulnerability and criticality of the area to be protected.
3) Desired level of protection.
4) Other security measures currently being used.
5) Availability of personnel.
6) Cost effectiveness.

There are several alarm and sensory devices applicable to perimeter intrusion detection. Given the proper environmental and operational conditions, various alarm system devices primarily used for internal purposes are adaptable to external usage. However, weather conditions, animals, birds, blowing objects, etc., must be considered prior to selecting a perimeter intrusion detection system. Systems currently being used to provide perimeter protection include: electromagnetic capacitance devices, narrow-beam radio frequency devices (operate similar to a photoelectric system), infrared or laser beam photoelectric systems, vibration/movement detection devices, and seismic devices.

The effectiveness of perimeter security measures is determined by the potential intruder's discernment of the fence, the CCTV camera, the intrusion detection system, and the security guard as psychological deterrents, physical impediments, and/or detection devices. If the security system is properly designed within the context of the environment and potential threats to its integrity, it will be effective in accomplishing the objectives for which it was intended.

Other Perimeter Barrier Applications

In many facilities it is often necessary to construct perimeter barriers around equipment or operations that are especially hazardous or unsafe. These barriers serve to provide limited or restricted access into areas defined as hazardous. A prime example of a "safety" barrier is that of a fence surrounding the immediate area of an electrical substation. Only those people who have both the need and the level of expertise to enter the substation should be allowed to do so.

Another and perhaps more obvious aspect of safety is derived from the perimeter barrier surrounding the total facility. Most industrial and manufacturing operations are not open to the general public to come and go as they please. The operations and equipment utilized are dangerous to the untrained and unaware nonemployee. The perimeter fence serves to deny free and unobstructed entry into what is essentially a hazardous environment. Freedom from interference by outsiders is assured and organizational liability for improper or inadequate safety measures is avoided.

Perimeter barriers are also utilized to complement and supplement fire prevention efforts. Areas containing very hazardous materials or processing operations can be separated from the larger facility, thereby allowing for more stringent control of fire regulations and standards. Mobile barriers are particularly useful when temporary conditions exist that are especially hazardous or unsafe. Concertina wire and other movable barriers can be utilized to block traffic lanes or isolate particular areas.

Crime Prevention Through Environmental Design

The environment to which one is exposed is fundamental in determining how one acts and perceives his surroundings. Thus, it is both natural and imperative that one should seek to understand its influence upon both crime and the fear of crime within society.

In 1974, a major program of Crime Prevention Through Environmental Design was launched by the National Institute of Law Enforcement and Criminal Justice. Residential, commercial, and school environments and the predatory, fear-producing crimes in each are the focus of this program.

The basis of CPTED to achieve security can be found throughout history. For example, moats and fortress walls were built around medieval cities to reduce external threats. Lighting programs have a precedent, too: in the seventeenth century, some 6,000 lanterns were installed on Paris streets as part of a crime reduction program.

The CPTED concept is focused upon the interaction between human behavior and the "built environment," including both natural and man-made elements. The physical design of an environment can facilitate surveillance and access control of an area and can aid in creating a sense of property awareness (territoriality.) Proper space definition through the design or utilization of natural and man-made barriers can:

1) Extend the area over which one feels a proprietary interest and responsibility so that his area overlaps that of other responsible citizens or entities.
2) Increase one's ability to perceive when his "territory" is potentially threatened and permit him to act on that perception.
3) Provide a potential offender or intruder with a perception that he is trespassing on someone else's domain, thereby deterring him from criminal behavior.

When possible, the CPTED approach emphasizes natural access control and surveillance created as a byproduct of the normal and routine use of the environment. It seeks to deter or prevent crimes and their attendant fears by careful design of the environment.

This chapter has addressed the planning, development and deployment of perimeter security. It should be remembered that perimeter security is the first line of defense against intrusion and the last line of defense against inventory shrinkage. The security manager should evaluate both the natural and man-made elements of a facility and build upon those features that would reduce vulnerability to security hazards. It is also essential that those responsible for planning and operating security programs be aware of new and important developments such as CPTED which might serve to refine and improve existing security efforts.

Discussion Questions

1. Briefly list and define the security reasons for utilizing perimeter barriers.
2. Identify the benefits from the utilization of gate or entrance security personnel.
3. Describe the benefits derived from creating a sense of property awareness (territoriality).
4. Discuss the circumstances where each type of perimeter fencing is appropriate.
5. What are some of the security considerations involved in the construction of a parking lot?

Notes

1. Carl E. Pope, *Crime-Specific Analysis: The Characteristics of Burglary Incidents*, Law Enforcement Assistance Administration, 1977, p. 28.
2. Raymond M. Mombossiee, *Industrial Security for Strikes, Riots and Disasters*, Charles C. Thomas, 1968, p. 88.

Chapter 5
Security Lighting

A good security program will ensure that the facility is secure at night as well as during the day. The most common method of equalizing security between day and night is the installation of protective lighting. Protective lighting not only enhances the security effort, but it also serves as a deterrent for potential criminal activity. This point was verified by a study of six areas in California by the National Criminal Justice Information and Statistics Service. The study showed that in 69 percent of the burglaries studied, the point of entry was not illuminated.[1] In the majority of cases, burglars selected victims on the basis of visibility.

The deterrent effects of protective lighting also aid the police in reducing crime. For example, in 1970 the city of Washington, D.C., initiated several street lighting projects using sodium vapor lights. The increased lighting contributed to marked decreases in burglaries and robberies for areas where lighting was installed. A number of cities are now experimenting with and implementing this program.[2]

With regard to law enforcement activities, street lighting has had excellent results in other cities.

> A dramatic instance of the effect of good lighting was the situation in New York City when street crimes of violence reached catastrophic heights. The police department flooded the crime-ridden area with plain clothesmen, uniformed foot patrolmen, and special squads with practically no success. Then the New York Police Department outlined 110 blocks where crime was most rampant, and the city installed new bright lights. In that newly lighted area, crimes of violence were cut in half; juvenile crime was cut by a third; and all crimes, including automobile thefts, were drastically reduced.[3]

It is interesting to note that, in this instance, protective lighting was much more effective in reducing crime than the deployment of additional manpower. If protective lighting possesses such rewards for the police, then it can be argued that similar benefits will be bestowed upon the security manager who effectively plans for and installs a protective lighting system. In addition to its effectiveness, protective lighting is inexpensive when properly deployed and may reduce the need for additional security forces. Therefore, protective lighting is an essential part of any security system.

Historical Aspects

The first street lighting systems were installed in Paris, France in 1558. Pitch-burning lanterns were placed on some of the main streets. An ordinance was also passed requiring citizens to keep lights burning in windows that fronted streets.[4]

In 1805 the National Light and Heat Company started using gas lighting in London, and two years later gas lamps were used to illuminate public streets. In 1872, street lights using electric filaments were used.[5]

Figure 5-1
Street Light Innovations.[6]

Date	Place	Light Source/Lamp
1558	Paris, France	Pitch-burning lanterns, followed by candle lanterns
1690	Boston, Massachusetts	Fire baskets
1807	London, England	Gaslights
1879	Cleveland, Ohio	Brush arc lamps
1905	Los Angeles, California	Incandescent
1935	Philadelphia, Pennsylvania	Mercury vapor
1937	San Francisco, California	Low-pressure sodium
1952	Detroit, Michigan	Fluorescent
1967	Several U.S. cities	High-pressure sodium

Today, even the smallest communities provide for street lighting in their budgets. As protective lighting in the public sector has developed, it has come to serve many functions. The National Evaluation Program on Street Lighting Projects outlines the varied uses:

Impact Objectives of Street Lighting Systems[7]

Security and Safety

- Prevent Crime
- Alleviate Fear of Crime
- Prevent Traffic (Vehicular and Pedestrian) Accidents

Community Character and Vitality

- Promote Social Interaction
- Promote Business and Industry

- Contribute to a Positive Nighttime Visual Image
- Provide a Pleasing Daytime Appearance
- Provide Inspiration for Community Spirit and Growth

Traffic Orientation and Identification

- Provide Visual Information for Vehicular and Pedestrian Traffic
- Facilitate and Direct Vehicular and Pedestrian Traffic Flow

While the above goals are obviously the concerns of law enforcement, they are also the concerns of the security manager since the same problems exist within his area of responsibility.

Planning Considerations

For the most part, protective lighting serves three distinct purposes. First, to the facility owner or manager, it serves to advertise his product or service during the evening hours. This frequently causes problems for the security manager, since he is more interested in security than aesthetic qualities. For example, the security manager would probably prefer flood lamps mounted near the roof, out of reach of potential intruders, and directed downward exposing a large area immediately adjacent to the structure. The business or facility manager, on the other hand, would probably prefer flood lights mounted in the ground, exposing the exterior of the building and possibly a sign for advertising purposes. When protective lighting is used for advertising purposes, it frequently confounds security efforts.

Secondly, protective lighting is used to facilitate pedestrian and vehicular traffic within a compound. For example, roads, entrances and exits, pathways and parking facilities should be lit during darkness. Again, a conflict may arise when the same lighting is used to promote safety and security. A pattern of lights adjacent to a structure used to provide visibility for vehicular traffic may not be the best arrangement to promote security for the structure itself.

Finally, security lighting is deployed to deter unauthorized entries and exits from the facility, and when it does not deter, it aids the subsequent discovery and apprehension of intruders. The commission of a crime includes three elements: the desire, the ability, and the opportunity. Protective lighting affects the desire and opportunity. It serves as a psychological barrier thus reducing a perpetrator's desire, and it reduces the opportunity by aiding apprehension.

Thus, protective lighting serves a number of purposes. In order to achieve the most advantageous use of lighting, a considerable amount of planning by facility managers, security managers, and engineers must take place.

Protective Perimeter Lighting

Protective perimeter lighting is an essential element of an integrated, complete physical security program. The application, placement, and level of security lighting depends on each specific location and structure to be protected. The type of perimeter, e.g., a fence line, a building wall, isolated, or semi-isolated, is a determining factor in the lighting system to be utilized. (See figure 5-2) Good perimeter lighting, wherever it might be located, is achieved by adequate, even light upon bordering areas, glaring lights in the eyes of potential intruders, and relatively little light on security personnel and their patrol routes or stationary posts. Protective security lighting should enable security personnel to observe activities around and inside the facility without being "on stage" themselves.

Figure 5-2
Perimeter Lighting Requirements.[10]

Type of area	Type of lighting	Width of lighted strip (ft)	
		Inside fence	Outside fence
Isolated perimeter	Glare	25	200
Isolated perimeter	Controlled	10	70
Semi-isolated perimeter	Controlled	10	70
Non-isolated perimeter	Controlled	20-30	30-40
Building face perimeter	Controlled	50 (total width from building face)	
Vehicle entrance	Controlled	50	50
Pedestrian entrance	Controlled	25	25
Railroad entrances	Controlled	50	50
Vital structures	Controlled	50 (total width from structure)	

Lighting used for perimeter security will not be the same as that used for illuminating streets, roadways, or work areas. While almost any level of lighting is helpful, proper levels of intensity, coverage, and placement are necessary to maximize the effectiveness and efficiency of lighting types and sources. Lighting units should be placed in positions that offer maximal coverage and security for the unit being used. For example, lighting for perimeter fence lines should be located inside the facility, directed outward, and located so as to achieve the needed level of coverage both inside and outside the perimeter.

As previously stated, lighting is inexpensive to maintain and can sometimes be employed to negate or reduce the need for more expensive security measures. However, the nature and substance of protective lighting is that it serves to deter potential intruders, and as such, cannot stand alone. It must supplement, and in turn, be supplemented by other security measures.

Visual Factors

In planning for an effective protective lighting layout, there are four visual factors which must be taken into consideration—size, brightness, contrast, and time.[8] The lighting layout depends on desired degree of security and the nature of the objects and environment being secured. Generally, larger objects require less light than smaller objects. The larger the object, the more light it will reflect, thus requiring less illumination. Brightness refers to the reflective ability of the object or structure. Light colors such as white reflect more light than dark colors such as black or brown. Thus, a building painted white would require less light than comparable buildings constructed of dark brick. Additionally, the texture of the objects under observation affects needed light intensity. Coarsely textured objects tend to diffuse light, whereas smooth surfaced objects tend to reflect light, reducing the need for higher intensity protective lighting. Contrast refers to the relative shapes and colors of objects under observation in relation to the total environment. If there is contrast between the objects being secured and the immediate environment, observation is much easier than if there was little relative contrast. Finally, time refers to the fact that greater illumination is required for areas that are visually complex or crowded, because this makes it harder to scan quickly or for extended periods. Open spaces, on the other hand, require less light because the observer has more time to observe and focus on seemingly foreign objects.

In planning protective lighting patterns, it is important to consider all four of these visual factors. During the planning stages, various data should be collected:[9]

1) Descriptions, characteristics, and specifications of the various incandescent, arc, and gaseous discharge lamps.
2) Lighting patterns of the various units.
3) Typical layouts showing the most efficient height and spacing of equipment.
4) Minimum protective lighting intensities required for various applications.

Lighting Terminology

When comparing various lighting systems, it is important that one be familiar with the terminology used in rating the effectiveness of various lamps. Some of the more commonly used terms are as follows:[11]

1) Candle power - One candle power is the amount of light emitted by one standard candle. This standard has been established by the National Bureau of Standards and is commonly used to rate various systems.
2) Foot candle - One foot candle equals one lumen of light per square foot of space. The density or intensity of illumination is measured in foot candles. The more intense the light, the higher the foot candle rating for the light.
3) Lumen - One lumen is the amount of light required to light an area of one square foot to one candle power. Most lamps are rated in lumens.
4) Brightness - Brightness refers to the ratio of illumination to that which is being observed. High brightness on certain backgrounds causes glare, and low brightness levels on some backgrounds makes observation difficult. Brightness, therefore, should not be too low or too high relative to the field of vision.

On a clear day during the middle of summer the sun supplies approximately 10,000 foot candles to earth. This intensity can be measured in clear openings such as fields or along the shore. At the same time, the amount of illumination under a shade tree is approximately 1,000 foot candles. At night, the average living room is reduced to approximately five foot candles.[12]

Types of Protective Lighting

Protective lighting is divided into four general categories: continuous lighting, standby lighting, movable lighting, and emergency lighting. The type of lighting selected will depend upon the nature of the security problem.

Continuous or stationary lighting is the most common type of protective lighting. Continuous lighting is the installation of a series of fixed luminaries so that a particular area is flooded with overlapping cones of light. There are two methods of deploying continuous lighting: glare projection and controlled lighting.

Glare projection lighting involves lights aimed directly at the observer so that a potential intruder's observation is impaired when attempting to look into the facility or structure. However, when using this method of lighting, particular caution must be taken to ensure that adjacent operations or traffic are not impaired. Glare projection lighting is a strong deterrent to potential violators since observation into the secured area is difficult, and a guard or watchman can be easily concealed in comparative darkness.

Glare projection lighting is extremely useful in illuminating perimeter barriers. Such a lighting configuration compounds the effects of the barrier. This method is also useful in lighting entrances or checkpoints since it increases the guard's visual powers while reducing the inturder's.

Flood lights which illuminate a wide horizontal area are the best source of lighting when glare projection is desired. The flood lights should be mounted on poles, along a roof line, or atop a wall or fence directed outward and downward. Exact placement of lighting depends on the degree of security and the nature of the immediate environment (figure 5-3).

Figure 5-3
Perimeter Protection by Glare Lighting.

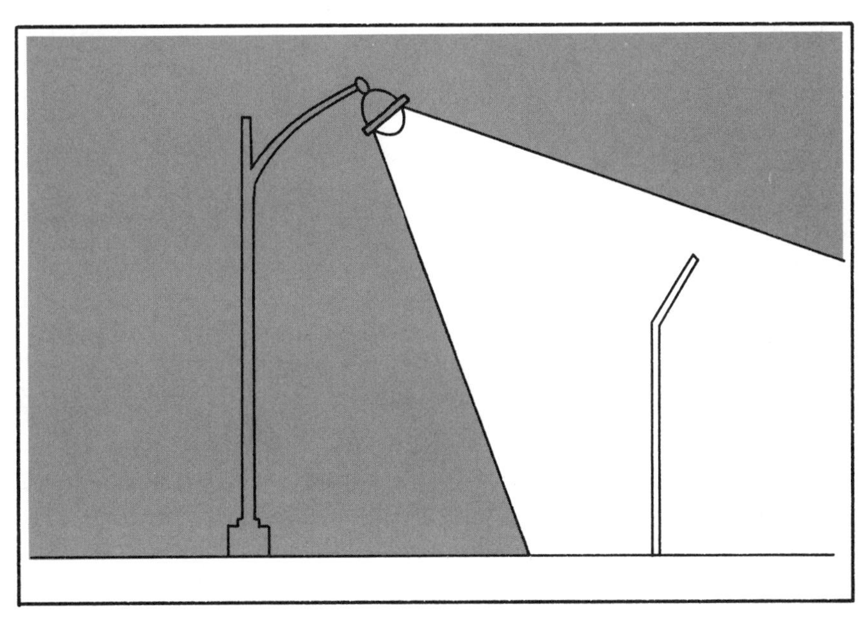

Controlled lighting is what the name implies—it is used to light a particular area in a controlled fashion. It is used to light facilities or areas which cannot or should not utilize glare projection lighting. Frequently, glare projection lighting cannot be used because it creates a dangerous situation for adjacent activities. For example, the structure may be adjacent to a highway, a parking lot, or where employees are working. In these cases, glare lighting may be dangerous or it may detract from the work efforts of the nearby employees, so some form of controlled lighting would be appropriate (figure 5-4).

Figure 5-4
Perimeter Protection by Controlled Lighting.

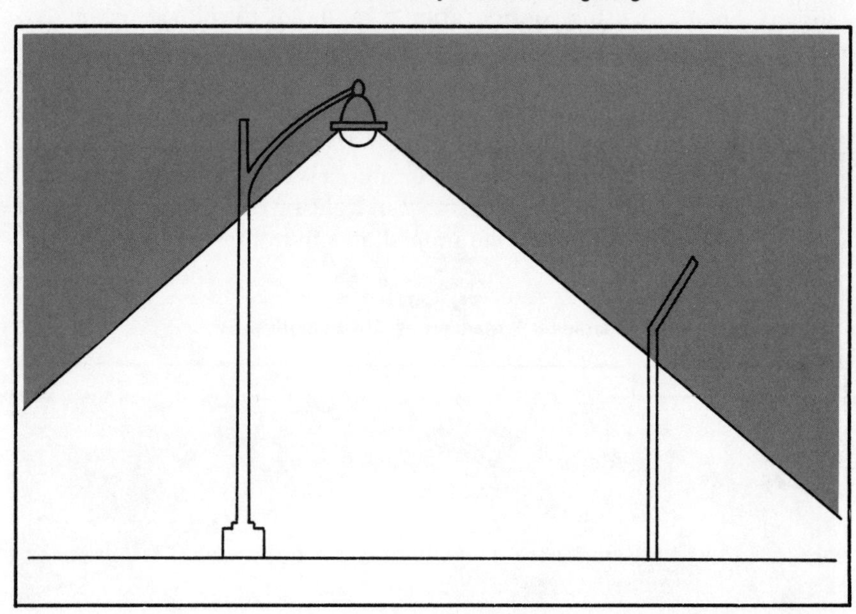

Controlled lighting uses lights mounted on poles, buildings, or fences and directed primarily downward, thus reducing the amount of glare. Of course, one problem with controlled lighting is that it often illuminates the facility being secured and any guard or patrol activities.

Standby lighting configurations are similar to continuous lighting except that standby lights are not continuously lit, but are manually or automatically turned on in specific, predetermined situations. The lighting may be activated when a suspicious activity is observed or when it is suspected that an intruder may be lurking about.

Movable lighting is a manually operated, mobile lighting system. Movable lighting should be available to supplement continuous or standby lighting. Additionally, movable lighting should be available to enhance security operations where security is not normally provided, for example, during an operation at an infrequently used rail loading dock.

Finally, emergency lighting refers to a system which, more or less, duplicates all of the above systems. It is generally used in situations where regular lighting is inoperable during power outages or emergencies. Such a system must include its own power source.

Types of Light Sources

The basic types of light sources most commonly used for security purposes are: incandescent or filament lamps and gaseous discharge lamps. Gaseous discharge lamps include mercury vapor lamps, sodium vapor lamps, metal halide lamps and fluorescent lamps. The selection of a particular type of lamp depends upon the nature of the security problem.

Figure 5-5 shows the historical development of the various types of lights:

Incandescent or filament lamps are common glass light bulbs where light is produced by the resistance of a filament to an electrical current. These bulbs are commonly used in homes and the work place. Their use is somewhat limited in security systems because of their low rated life and their lower lumens per electrical watt rating.

Mercury vapor lamps are more efficient than incandescent lamps. They emit a blue-green light and are used to light both interior and exterior work areas, because the light color is not as distracting to the human eye as is the light produced by other types of gaseous lamps.

Sodium vapor lamps are similar in construction to the mercury vapor lamps, and are the most efficient lamps in use today. As noted in the above table, high-pressure sodium lamps produce up to 140 lumens of light per watt and have a rated life of 15,000 hours. One problem with these lamps is that they emit a golden yellow light which is somewhat harsh and so are not appropriate for work areas. However, these lamps provide excellent security lighting, especially around perimeter barriers because of their efficiency. Today many streets in high crime areas use sodium vapor lamps.

Metal halide lamps are similar in nature to sodium vapor lamps. They emit a harsh yellow light which is extremely distracting. They contain sodium, thalium, indium and mercury.

Fluorescent lamps have a rated life of up to 14,000 hours, but they do not produce as much light as some of the gaseous discharge lamps. They are appropriate for work areas since the light is not as distracting as some of the gaseous discharge lamps.

Although gaseous discharge lamps are more efficient, one problem which limits their application for security purposes is the amount of time it takes them to light up. On the average it takes a gaseous discharge lamp approximately four minutes to warm up and become fully operative. If power was interrupted, it could cause the security system to become ineffective for a short period. Incandescent and fluorescent lamps do not have this problem.

The selection of a particular type of light source is dependent upon a number of conditions. Some of these conditions are:[14]

Figure 5-5
Historical Development of Street Lighting.[13]

Lamp Description	Date	Rated Life for Street Lighting Service	Initial Lumens Per Watt
Arc			
Open carbon-arc	1879	Daily trimming	—
Enclosed arc	1893	Weekly trimming	4-7
Flaming arc			
Open	—	12 hours	8.5 (d-c multiple)
Enclosed	—	100 hours	19 (a-c series)
Magnetite (d-c series "luminous arc")	1904	100-350 hours	10-20
Filament			
Carbonized bamboo	1879	—	2
Carbonized cellulose	1891	—	3
Metallized (gem)	1905	—	4
Tantalum (d-c multiple circuit)	—	—	5
Tungsten (brittle)	1907	—	—
Drawn tungsten	1911	—	9
	1913	—	10
Mazda C (gas-filled)	1930	—	14-20
	1915	1,350 hours	10-20
	1950	2,000 hours	16-21
		3,000 hours	16-20
Mercury Vapor			
Cooper-Hewitt	1901	Indefinite	13
H33-1CD/E	1947	3,000 hours	50
H33-1CD/E	1952	5,000 hours	50
H33-1CD/E	1966	16,000 hours	51
H36-15GV	1966	16,000 hours	56.5
Low-Pressure Sodium			
NA 4 (10,000 lumen)	1934	1,350 hours	50
NA 9 (10,000 lumen)	1935	2,000 hours	56
	1952	4,000 hours	58
	1975	—	180
Fluorescent			
F100T12/CW/RS	1952	7,500 hours	66
F100T12/CW/RS	1966	10,000 hours	71
F72PG17/CW	1966	14,000 hours	68
F72T10/CW	1966	9,000 hours	63
High-Pressure Sodium			
	1965	6,000 hours	Over 100
	1975	15,000 hours	140

1) Cleaning and replacement of lamps and luminaries, particularly with respect to the costs and means (e.g., ladders, mechanical "brackets," etc.) required and available.
2) The advisability of including manual and remote controls, mercury or photoelectric controls.
3) The effects of local weather conditions on various types of lamps and luminaries.
4) Fluctuating or erratic voltages in the primary power source.
5) The requirement for grounding of fixtures and the use of a common ground on an entire line to provide a stable ground potential.

Types of Lighting Equipment

Basically, there are four types of lighting equipment available: flood lights, street lights, fresnel units and searchlights. The usage of a particular type is dependent upon the particular security problem.

Floodlights are most amenable to security needs because they project light in a concentrated beam. Thus, they can be used to light a particular point or area. Floodlights are manufactured with a variety of beam widths, enabling one to select the appropriate light to meet the needs of the task at hand. Since floodlights emit a directed beam, they are appropriate for use in instances which call for glare projection lighting, i.e., the illumination of boundaries, buildings, or fences.

Street lights produce a diffused light rather than a directional beam. They geneally produce little glare and are appropriate for use in controlled lighting situations. They are commonly used to light parking lots, thoroughfares, facility entrances and boundary perimeters where glare is undesirable. Additionally, because of their efficiency, they can be deployed to light large areas at a minimal cost.

Fresnel units emit a fan-shaped beam of light, covering approximately 180 degrees horizontally and 15 to 30 degrees vertically. Fresnel units are the most effective units when glare lighting is desired because of their projection pattern. Additionally, when they are mounted on a tall pole and directed downward they are effective in lighting areas between buildings. For the most part, they are used to light perimeters since little light is lost vertically.

Searchlights are lighting units which produce a highly focused beam of light. They can be stationary or mobile units depending on their application. Because searchlights can be aimed in all directions and are often mobile, they are commonly used to complement existing lighting systems.

Designing a Lighting System

Selection of a lighting system and fixtures is dependent upon the purpose of the system and the environment. For example, if a perimeter fence is at least 100 feet from structures or work areas and there is a clear zone on the outside of the perimeter, then either controlled illumination or glare projection techniques are appropriate. Both techniques would adequately illuminate the barrier. If either posted guards or patrol units are used, they should be posted outside the illuminated areas. The same principles apply to semi-isolated and non-isolated perimeters. Lighting should be deployed to provide full illumination of the barrier, but it must not interfere with guard activities, traffic or other businesses or activities inside or outside the perimeter.

When illuminating buildings special care must be taken to ensure that activities are not inhibited. If the structure facade contains no windows, then glare projection from ground units would be appropriate since they would provide maximum exposure of the building surface. When this type of lighting is used, special care should be taken to ensure that intruders do not sabotage individual units. If guards are posted or there are windows or other openings that reveal work activities, then glare projection from roof-mounted units is more appropriate. Additionally, glare projection lighting is not appropriate at vehicular or personnel entrances since the glare would create an unsafe condition. Here, controlled lighting is most appropriate. Care must also be taken to ensure that the lighting does not leave shadows which would enable intruders to hide.

Entrances for pedestrian and vehicular traffic should be lit using controlled lighting. The lighting should be intense enough to enable guards to recognize persons and to examine credentials and other papers. If the entry point has a gate house, the level of illumination should be lower so that those approaching will have difficulty in discerning the activities of the guard. This would aid in the deterrent effects of the perimeter barrier.

All work areas within the compound, especially those where materials and merchandise are being loaded and unloaded, should be illuminated with some form of continuous lighting. If the work area and the area immediate to the work area are illuminated, it will reduce the probability of inventory shrinkage due to employees hiding the property in the immediate area.

In summary, the security manager must consider three factors in planning the protective lighting system: security needs, costs, and operations safety. Each security situation must be thoroughly examined with these factors in mind. Only after such an analysis can the best lighting configuration be devised and deployed.

Discussion Questions

1. Outline and discuss the security objectives of street lighting systems.
2. Briefly discuss the three purposes of protective lighting.
3. The commission of a crime includes three elements. List these elements and discuss the role of protective lighting in diminishing crime.
4. List and describe the four general categories of protective lighting.
5. Discuss some of the conditions which should be taken into consideration when selecting a light source.

Notes

1. The Merritt Company, *Protection of Assets Bulletin*, No. 1, January, 1978.
2. B. D. Colen, "D. C. Lights the Way in Fighting Crime," *The Washington Post*, February 7, 1971, Sec. B.
3. Larry Vardell, "Report to the National Crime Prevention Institute," *Lighting for Crime Prevention*, Dade County Police and Safety Department (no date).
4. James M. Tien, Vincent F. O'Donnell and Arnold Barnett, *Street Lighting Projects*, National Institute of Law Enforcement and Criminal Justice, Washington, D.C., 1979, p. 3.
5. Ibid, Vardell, p. 1.
6. Ibid, Tien, et al., p. 5.
7. Ibid, Tien, et al., p. 3.
8. Richard J. Healy, *Design for Security*, John Wiley and Sons, 1968, pp. 136-139.
9. *Physical Security Manual FM-19-30*, Department of the Army, 1979, p. 91.
10. Ibid, pp. 4-11.
11. *Security Manual*, pp. 4-12.
12. Healy, p. 140.
13. Ibid, Tien, et al., p. 5.
14. *Security Manual*, pp. 4-11.

Chapter 6
Locks

The physical security of any property or facility relies very heavily upon locking devices. These devices vary greatly in appearance, as they do in function and application. A lock, regardless of its type, is primarily a delaying device. The degree of delay presented by the lock is dependent upon its quality of construction and installation, and the skill of the would-be intruder. Often the locking device is the first line of defense, whether it is on the perimeter fence line, a door to the facility, or an interior office.

History of Locks

Archaeological digs have uncovered Egyptian pin locks that date back approximately 3,000 years. These ancient locks were wooden and required keys so large that they were carried over the shoulder. Although molded from wood and quite large by today's standards, the ancient locking devices of the Egyptians and the Chinese utilized the same elements of alignment and positioning of component parts that are basic to their modern counterparts today.

Warded Locks

Warded locks, the simplist of which has only three moving parts, i.e., the bolt, an arm that moves the bolt, and the key that activates the bolt area were in use as early as the first century B.C. Warded locks of this same basic design can be found on the doors of many homes built prior to World War I and on cheap, low quality padlocks. (figure 6-1)

Lever Tumbler

The lever tumbler lock came into use somewhere around the turn of the nineteenth century. This device added one more step in the degree of security and protection afforded by locks. The introduction of the lever tumblers, movable pieces of metal between the key and the lock bolt, resulted in a locking device that was more complex and harder to pick or force open.

Figure 6-1
Warded Lock.

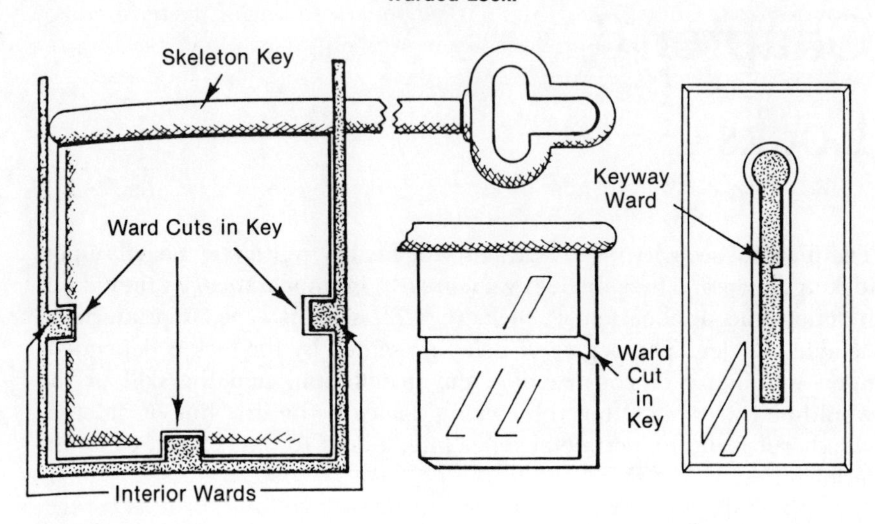

Wafer Lock

Wafer locks, often referred to as disc-tumbler locks, were developed after the advent of the lever tumbler lock and added yet another step in lock sophistication. Flat spring-loaded metal discs in the plug of the lock bind the core of the cylinder shell to the cylinder housing. When the proper key is inserted into the keyway the disc tumblers align and withdraw their protruding parts, allowing the plug to be turned. Today wafer locks can be found on automobiles, desks, cabinets, padlocks, etc.

Pin Tumbler Lock

In the 1850's, Linus Yale, Jr., developed what is known today as the pin tumbler lock. Based upon the ancient principles established by the Egyptians, Yale constructed an inner lock mechanism that is characterized by tumblers in the form of metal pins which rest vertically inside individual chambers housed in the cylinder shell. Today, the pin tumbler lock has become the most common type of locking device used for protection and security purposes.

More recent developments and innovations in locking devices include various types of electromechanical and electronic mechanisms. Many of these devices are integrated into systems of both identification and access control. Such systems can be programmed to permit or deny entry, and at the same time provide a record and identification of the user as to time and movement.

Lock Terminology

A prerequisite of lock security is a basic understanding of the terminology of lock security. The following terms are helpful in understanding the more common locks, locking devices, keys, functions and lock features.

Astragal—A molding to cover the opening (gap) between two meeting doors.

Barrel key—A cylindrical, hollow key with a projecting bit. The hollow end fits over and turns around a post in the keyhole.

Bit key—A key with a bit projecting from a round shank. Similar to the barrel key but with a solid rather than hollow shank.

Blank—An uncut or unfinished key.

Bolt—The part of the lock which is moved into a locked or unlocked position.

Bottom pins—The lowermost pins of a pin tumbler cylinder.

Bow—The handle or head of a key.

Cam—The part of a lock which activates the bolt as the key is turned.

Card operated locks—Electric or electromagnetic locking devices operated by coded cards serving as keys.

Code operated locks—Combination type locks in which no keys are used and instead are operated by pressing a series of numbered buttons in the proper sequence.

Combination—The arrangement of numbers to which a combination lock is set, or the arrangement of cuts on a key.

Cremone bolt—A vertical throw lock which locks the door or the sash into the frame at the top and bottom.

Cuts—The indentions made in a key to make it fit the tumblers of a lock.

Cylinder guard—A covering or device used to protect the cylinder of a lock.

Cylinder housing—The external case of a lock cylinder (also called the shell).

Cylindrical lock—A lock set having the cylinder(s) in the knob (also known as lock-in-the-knob).

Deadbolt—A lock bolt having no spring action and which becomes locked against end pressure when projected.

Dead locking latch—A spring bolt with an anti-shim device, which prevents the latch from being retracted by pressure applied to it.

Disc tumbler—A double-acting, spring-loaded flat plate designed to slide in slots in the cylinder plug.

Double bitted key—A key having cuts on two sides.

Driver—The uppermost pin in a pin tumbler lock.

Electromagnetic locks—Devices holding a door closed by electronically-induced magnetism.

Hasp—A fastening device consisting of a metal loop and a slotted hinged plate.

Header—Top cross member of a door frame.

Heel of a padlock—The stationary end of the shackle on a padlock.

Jamb—The vertical member(s) of a door or window frame.

Key—An instrument for operating a lock.

Keyhole—The opening in a lock to receive a key.

Latch—A device that secures or attaches but does not lock.

Lever tumbler—A flat piece of metal made to fit straight cuts in appropriate keys.

Lock—A device for fastening or engaging two or more objects which includes a means of manipulating the device into a locked or unlocked position.

Locking dog—That part of a padlock which engages the shackle and holds it in a locked position.

Locksmith—A person engaged in selling, installing, repairing , modifying and designing locking devices and keying systems.

Master pin tumbler—Cylinder pins which are usually flat on both ends; used to set a lock to accept more than one key.

Mortise lock—A lock with a bolt made for installing in a cavity (hole) cut in the edge of a door.

Padlock—A portable lock with a hinged or sliding shackle, normally used with a hasp.

Pin tumbler springs—Coil compression springs placed above or behind the driver pins in a pin tumbler lock.

Pin tumblers—Important parts of a pin tumbler cylinder, denoted as bottom pins, master pins, and drivers.

Plug—Round core of the lock cylinder which receives the key and rotates when the key is turned.

Retractor—The part of a lock which is attached to the bolt and moves to an unlocked position.

Rim lock—A lock designed to fit on the surface of a door.

Shackle—The hinge or sliding part of a padlock.

Shearline—The area between the housing and plug which is normally obstructed by the tumblers and becomes unobstructed by use of the proper key, allowing the plug to rotate.

Shoulder—The projection(s) on a key between the bow and the blade which prevents the key from passing too far into the cylinder.

Shell—The external case of a lock cylinder without the plug (also called housing).

Skeleton key—(See Warded key).

Spring bolt—A spring bolt retracts upon contact with the lip of the strike and then extends into the hole of the strike securing the door in a closed position.

Strike—A metal plate installed on or in a door jamb to receive the bolt.

Throw—The outward movement of a bolt; the distance it travels.

Wafer lock—A locking device utilizing a flat metal wafer tumbler.

Ward—An obstruction which prevents the wrong key from entering or turning the lock.

Warded key—One used in warded locks which will bypass obstructions in the keyway (often called a skeleton key).

Common Locking Devices

Locking devices can provide varying degrees and manners of security. A lock will go a long way toward discouraging burglars, thieves, or other would-be criminals; however, it must be the right lock selected on the basis of its use in conjunction with environmental activities and other security hardware.

The following is a discussion of various locking devices that are easily obtainable and are used in homes, businesses, and industry. First, however, it is important to point out that a locking device can be designed to operate in a specific manner, yet at the same time be any of a number of different types. For example, the single cylinder locking device described below can be a lever lock, a wafer lock, or a pin tumbler lock, or it can be a rim lock, a mortise lock or a cylindrical lock. Thus, the security function of the lock is dependent upon the type of locking device, and the desired level of security is dependent upon the kind of locking mechanism employed.

Single Cylinder Locking Devices

These locking devices are installed in doors or placed on other objects that must be secured from only one side. They require a key to open them from one side. The most likely application of a single cylinder locking device would be on a solid door far enough from glass panels or windows so that an intruder cannot break the glass, reach in and open the door from the other side. Most locking devices of this type have a thumb turn on the inside of the door. This permits easy locking or unlocking of the door, which is sometimes important for reasons of safety or quick exit.

Double Cylinder Locking Devices

This type of locking device is installed on doors that must be secured from both sides. A key is required to open or lock the door from either side. A door with glass panels or one next to glass panels would likely be fitted with a double cylinder locking device. Such locking devices may not be feasible for use in schools, hospitals, fire exits, etc., where for reasons of safety their use would be prohibited.

Emergency Exit Locking Devices

These devices allow for quick exit without use of a key, usually by means of a horizontal "panic bar." The device locks the door against entry and in many instances no external hardware is apparent at all. Frequently these emergency exits have an alarm device that sounds when the exit is used (figure 6-2).

Figure 6-2
Emergency Exit Locking Device with Alarm Unit.

Electric Locking Devices

Electric locking devices respond to an electric current which releases the strike. They are utilized more for remote operation and convenience than for a high degree of security.

Recording Devices

While not a locking device within itself, it is a feature that can be incorporated into most locking devices, either mechanically or electrically, to provide for a record of door use by time of day and/or by key used.

Vertical Throw Devices

Several variations are available in vertical throw locking devices. While a vertical throw bolt can be found in a rim lock, single cylinder lock, double cylinder lock, etc., there are variations that do not require a key and can only be opened or locked from the secured side of the door. An example of this is the "police lock" which uses an angled bar which fits into a receptacle in the floor and is secured to the door at the other end. Another variation is the vertical bolt, which can be installed in the floor or on the bottom of the door in a recessed position, and is pushed down into a floor well to prevent the door from opening (figure 6-3).

Figure 6-3
Vertical Throw Cremone Bolt.

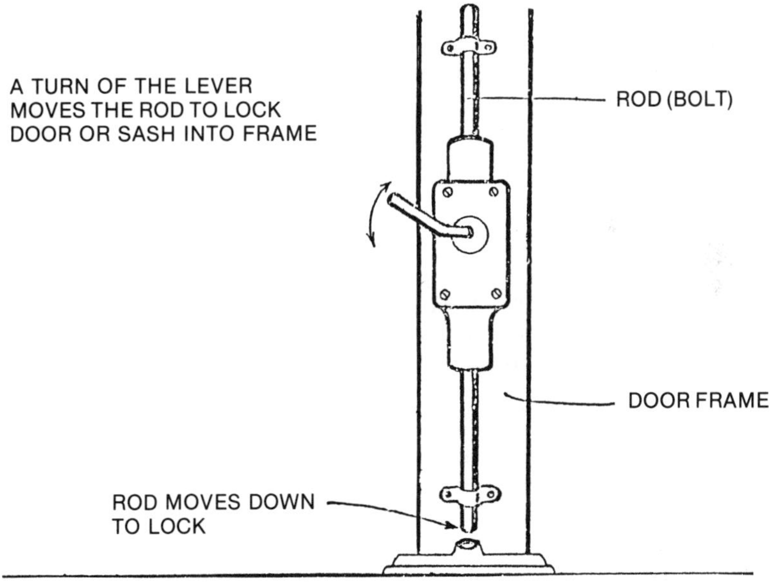

Sequence Locking Devices

Sequence locking devices are used to ensure that doors are closed and locked in a predetermined order. Each door is locked in its sequence and no door can be locked until its designated predecessor has been locked. This locking system prevents the forgotten unlocked door.

Keying Systems

A key is the standard method of accomplishing entry through a locked door and the normal way of locking it. Most key-operated locks are made to accept only one key which has been specifically designed and cut to fit it. Of course, a lock of any function, quality, or effectiveness, is worthless if keys are not available to those who need them. Keys and keying systems are generally divided into change keys, maison keys, control keys, submaster keys, master keys, and grand master keys.

Change Key — The standard type of key that fits a single lock within a master key system or to any other single lock unnumbered by such a system. However, be aware that numerous locks can be "keyed alike" to accept only one key.

Maison Key — A type of submaster key system very common in apartment houses and office buildings. Tenants are given a single key which operates both their apartment or office door and the main entrance door lock. This is done by using a lock having only two or three pin tumblers with many segments in each tumbler. The more tenants, the more shearlines there must be in the lock at the main entrance. This is a very insecure system of keying. A better and more secure practice would be to provide each tenant with one key for entrance to the building and a second key for entrance into the individual office or apartment.

Master Key Systems

The process of master keying consists of splitting the bottom pin into two or more segments so that keys of different combinations (cuts) will raise bottom pins and master pins to a shearline (figure 6-4). As the

Figure 6-4
Change Key and Master Key Illustrate Different Shearlines and Pin Alignment.

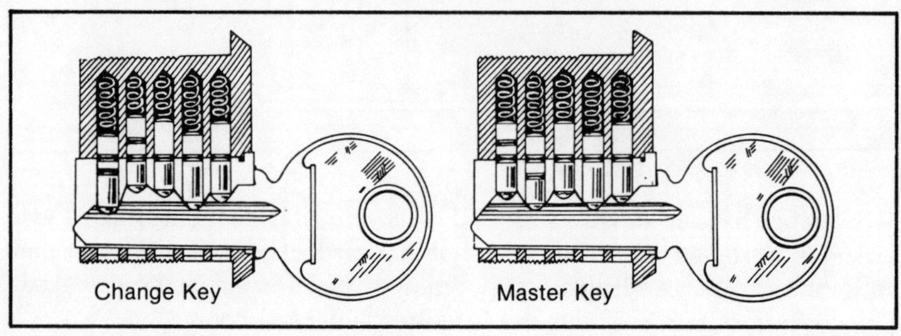

number of locks in a master key system increases, and as the progressive stages of master keying increase, so does the number of master pins and shearlines. With each increase there is a resulting decrease in security, because the chances of arriving at or finding a shearline for each pin by picking is increased sharply. There can be progressive stages of master keying. The following keys are found in a simple master key system.

Submaster—This is a key that will open all locks of a particular area or grouping within a given facility. The locks may be those of one floor of a multi-storied building, a particular operation such as administration, or even one building out of several.

Master Key—The master key will open all the locks in the facility that are incorporated into the master key system.

Grand Master Key—This key will open every lock in a keying system involving two or more master key groups.

Control Key—A control key is used primarily for maintenance or replacement purposes. It is cut in such a way that it operates to remove the core from the housing. Another core which requires a different key can then be inserted in a matter of seconds. This method is particularly useful when security of the facility has been decreased by lost or missing keys (figure 6-5).

Figure 6-5
Typical Breakdown of a Keying System Used In Industry.

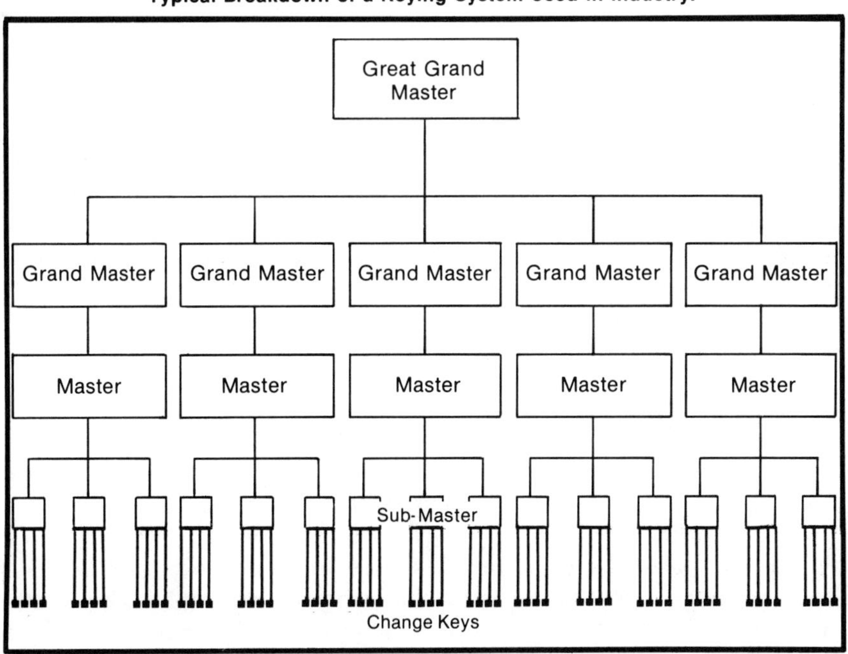

Key Control

A system must be present which accounts for and controls every key and every lock. Thus, responsibility and authority must be given to someone (preferably the security department) to maintain records, provide for a key depository, control issuance and retrieval, and investigate any misuse or loss of facility keys and locks.

Lock and Key Records

A written record and log should be maintained on all keys and locks. The issuance of keys should be controlled to provide keys only to those persons who have been shown to have a need for keys which would allow them entry into an area of the facility. When a key is issued, the record should indicate the key number, the name of the person to whom it is assigned, his position within the company, the date of issuance, and any other relevant data which might be beneficial.

A log should be kept of maintenance and repairs on locks, lost keys and actions taken to remedy any problems detrimental to lock and key security. All keys should be identified and secured in a high security key cabinet. All unissued and duplicate keys should be protected in this manner. As a general rule, the fewer keys issued, the more effective the security control.

Key Depository

It would be best if no keys to the facility ever left the premises. Though this is often not feasible or appropriate, the closer this goal can be approached the greater the degree of security a facility has over controlling access to its property. Keys taken off the property can be duplicated, lost or used in some other way to compromise the security for the facility. An ideal method would be for all employees who were issued keys each day to turn in or deposit the keys with security personnel at quitting time. A log of daily issuance and return would ensure that all keys were issued properly and accounted for.

Master Key Control

Obviously, master keys must be treated with greater care and security than change keys. The loss of a master key can threaten the entire keying system. A primary rule of key security is to minimize the number of keys given out, particularly master keys. Keys must not be issued for convenience nor should they be issued on the basis of an employee's position. Indiscriminate issuing of keys is little improvement over having no locks at all.

Master and submaster keys should not be marked or inscribed in any way that would identify them as master keys. A coding system for purposes of internal identification should be developed that would be known only by the necessary personnel. Whenever a key is issued to an employee, a lock becomes vulnerable to being compromised through the theft, loss, improper use, or duplication of the key. The loss or theft of one master key can result in the considerable cost of rekeying, and possibly losses of property.

Combination Locks

The combination locking device, provided it is of good quality and installed properly, generally affords a greater degree of security than most key operated locking devices. Commonly found on safes, vaults, high security storage cabinets, and high security padlocks, dial-type combination locks do not require keys to operate the lock mechanism. Generally, the integrity and security of a combination lock can be more effectively maintained than a key-operated lock, though combination locks are capable of being compromised if the combination should fall into the wrong hands or if the storage unit is improperly used. The combination must be subject to effective security procedures and controls. The combination must be restricted to an absolute minimum of personnel. Any written record of the combination must be afforded the highest security and if feasible no such record should even be kept. Combinations should be changed periodically as a matter of procedure and changed after the termination or transfer of any employee who knows the combination or worked in close proximity to the storage unit. Of course, a change should be made at any other time there is a suspicion that the security of the combination has been compromised.

Padlocks

Padlocks have a variety of security applications: perimeter fenceline gates, building doors, storage areas, equipment lockouts, employee lockers and tool chests. They can be incorporated into a master key system or be operated by a change key. A key-operated padlock has three basic parts: the key, the casing, and the shackle. The casing houses the internal locking mechanism and the keyway. The shackle is the locking or holding part of the padlock (figure 6-6).

The internal locking mechanism of a padlock can be of the warded, lever, wafer, or pin-tumbler type. It is generally agreed that a pin tumbler padlock having at least five pins in the cylinder offers the greatest degree of security from manipulation. However, the secure padlock must also be constructed of case hardened metal that is resistant to cutting and hammering.

Like other locking devices, a padlock is effective only when the surface on which it is installed is of solid construction. Care must also be taken that the hasp be case hardened, installed properly, and does not expose the mounting screws or bolts when in the locked position.

The procedures for the control and security of keys or combinations to padlocks should be incorporated into the total key control program. Routine inspection and maintenance of padlocks is important as they are often exposed to the environment and isolated on seldom-used perimeter gates. Security guards on regular patrol should routinely inspect such padlocks and the attached fastening device for evidence of compromise or deterioration.

Figure 6-6
Parts of a Padlock.

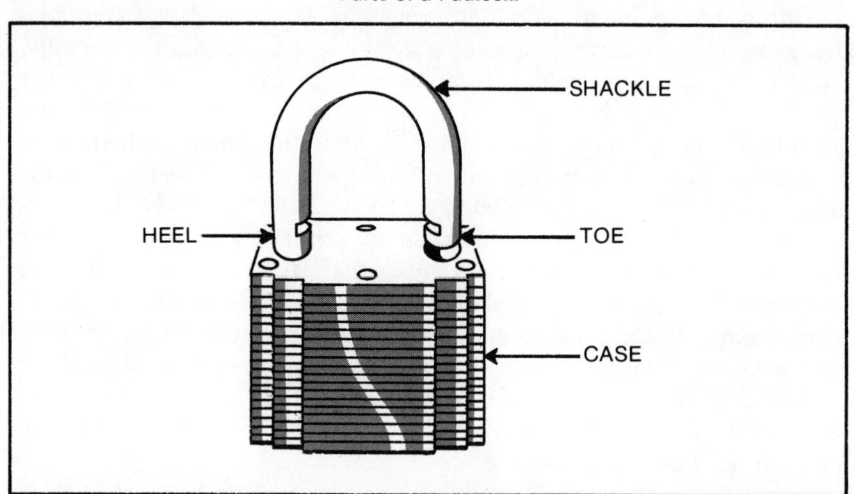

The Pin Tumbler Lock

Most locks manufactured today utilize the pin tumbler cylinder system. Pin tumbler cylinders are constructed of five basic components, the most visible part being the shell or housing encompassing the entire cylinder. This shell contains three to seven cylindrical chambers drilled from the top down through to the smaller circumference opening near the bottom. This opening is filled with a cylinder plug, which has an equal number of cylindrical chambers aligned directly with those drilled in the shell. This plug is retained in position by means of a cam or tailpiece normally attached with two screws from the back. This cam is the activating lever for the lock itself. In order for the cylinder to function as a security device, a set of coil springs and pin tumblers are placed within each chamber, from the top of the shell down. They would be sequenced with a coil spring applying pressure to the driver pin. Normally this driver pin is flat on both ends; however, various configurations such as a mushroom shape or cone shape are used to provide additional pick resistance. This driver pin is pressed against the top of the bottom or combination pin which is flat on one end and tapered to a point on the bottom side. The point at which the driver pin meets the bottom pin is known as the shearline.

When the proper key is inserted within the lock, the springs force bottom pins downward into the cuts in the top of the key. The depth of each cut in the key is proportional to the depth of the bottom pin, which means the top of each bottom pin will align itself at the breaking point between the plug and the cylinder shell, allowing the key to turn the plug without resistance (figure 6-7).

The versatility and security provided by the pin tumbler cylinder make it applicable to the simplest or most complex keying system.

Figure 6-7
Standard Pin Tumbler Lock Mechanism Without Key.

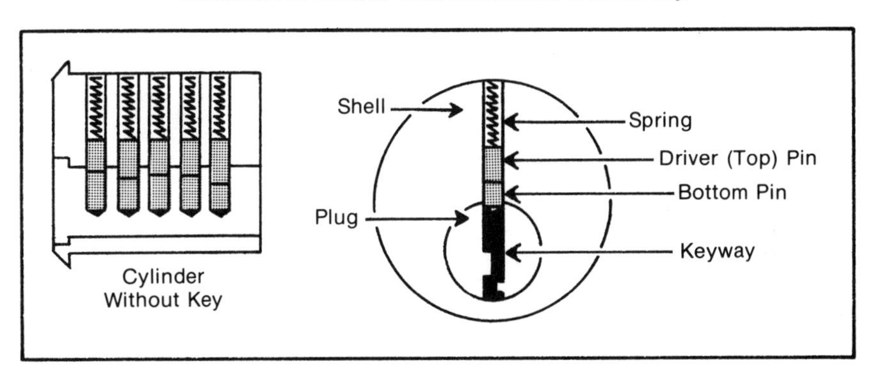

Types of Lock Bolts and Latches

The latch or bolt is that part of the lock which secures the door, window, gate, or other movable object to a stationary fixture. All bolts or latches, regardless of type, serve the same function. Bolts and latches can operate in two directions: slide horizontally across the door into a strike or slide vertically into a frame or floor receptacle at the top or bottom of the door.

The Deadbolt

A deadbolt does not contain a spring, and so must be manually moved into its locking position within the strike by turning a key or thumb turn.

It does not automatically move into the strike and secure the door when it is shut, and when the bolt is thrown into a locked position the door cannot be closed until the bolt is withdrawn. Deadbolts are usually rectangular in shape and provide a greater degree of security than latch type devices. Quality of construction and the throw of the bolt are important determinants in the level of security provided by deadbolt mechanisms. The greater the outward movement of the bolt, whether vertical or horizontal, the greater the degree of security.

The Spring-Loaded Latch

There are two kinds of latches: the spring-loaded latchbolt and spring-loaded deadlatch. The spring-loaded latchbolt provides a minimum level of security, because the latch can be withdrawn from the strike whenever force is applied directly to the latch itself. When the door is closed, the spring bolt automatically retracts upon contact with the strike and then extends into the hole of the strike, securing the door in a closed position. Spring-loaded latchbolts have a beveled end to allow this depression of the latch to occur. They are withdrawn from the strike by a turn of a key, thumb turn, or door knob.

The spring-loaded deadlatch operates in the same manner as the spring-loaded latchbolt except for an extra latch piece (bar) located on the side of the latchbolt, which when depressed against the edge of the strike, locks the deadlatch into position. These are also called deadlocking latches or spring bolts with an anti-shim device.

Lock Violations and Physical Assaults

Of course, the most common method of gaining entry through a locked door or other fixture is to use a key. This is the expected and proper way to gain legal entry. However, numerous methods and techniques are utilized by would-be intruders to defeat locks. Surreptitious violations and physical assaults include picking, jimmying, prying, jacking, smashing, carding, and drilling.

Picking is a method by which the lock's tumblers are manipulated through the keyhole with small tools made for this purpose. Picking a cylinder is usually done in one of two ways. One method is using a small tool called a tension wrench, which puts tension on the plug and hence on the pins which are impeding the rotation of the plug. With this tension tightening the pins, the pick is inserted to raise the pins, one by one, to the right level until the shearline is obtained for all the pins (figure 6-8). Another method is "raking," where a thin metal tool is inserted into and then quickly pulled out of the keyway, jostling the pins into position.

Carding or "loiding" is the action of slipping or shimming a spring bolt

Figure 6-8
Tools Found in Lockpicking "Kit".

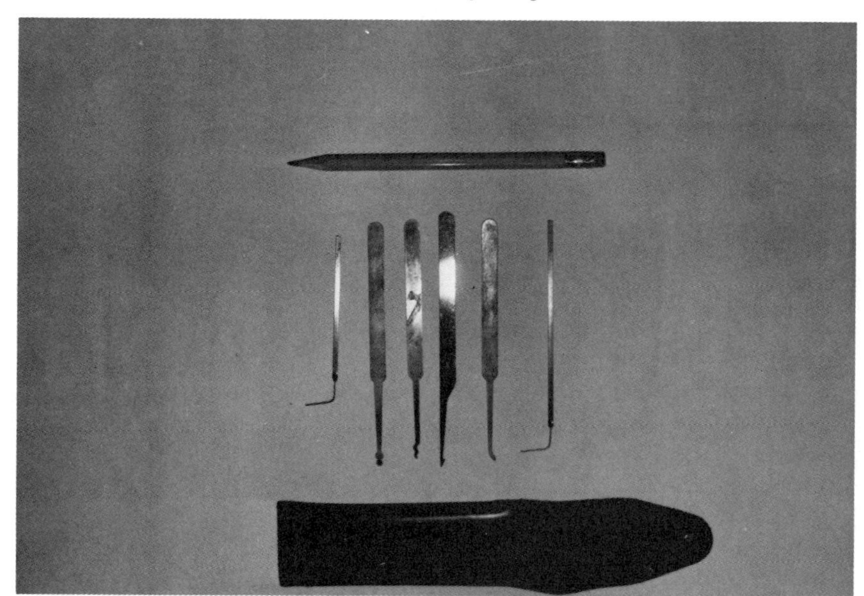

with a piece of celluloid. The spring-loaded latch which does not resist end pressure is particularly susceptible to carding.

Jacking is accomplished by placing an ordinary car jack horizontally between the door jambs and applying pressure. A poorly constructed door or one not sufficiently fortified will spread, pulling the bolt or latch out of the strike and allowing the door to open.

Jimmying and prying are accomplished by using tools such as crow-bars, smaller jimmy bars, large screwdrivers, or other metal tools to pry the door away from its frame or break the locking mechanism. Vertical locking devices will usually offer greater resistance to jimmying than horizontal devices (figure 6-9).

Smashing is simply a physical assault on the door. It can range from total destruction of the door to the breaking of a glass panel in the door to gain access to a thumb turn. Various tools can be used to smash a door. This technique, while not common, occurs when a building or other structure is located in an area remote enough that the noise created by the smashing is not a deterrent to breaking in.

Another method of bypassing a locked door is to remove the hinge pins, if they are on the outside. If at all possible, hinges should be located on the interior side of the door, but if they are exposed they should be welded in position or otherwise made nonremovable.

Figure 6-9
Jimmy Resistant Rim Lock.

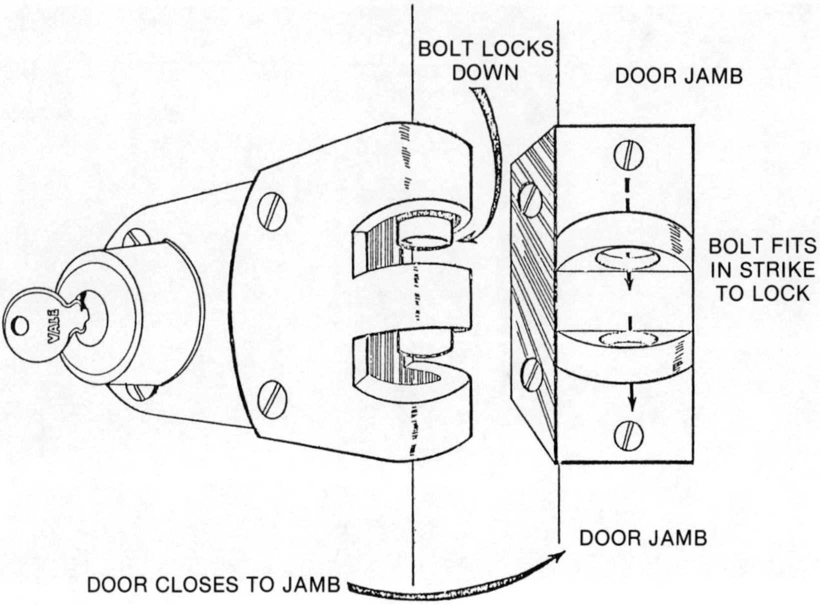

Basic Door Locks

Crime statistics indicate that approximately fifty per cent of all illegal entries occur through a door. It is axiomatic then that doors should be secured against surreptitious violations and physical assaults. The same is also true for windows which are the point of entry in another forty per cent of illegal entries.

Obviously, windows and doors are chosen as points of entry because they are the most common openings in buildings, yet it is not so apparent why one building is broken into as opposed to another. Anyone desiring illegal entry into a building or facility, in many, if not most cases, bases his choice on the perceived ease of entry and the level of risks involved. The perceived ease of entry has a direct relationship to the level of risk taking, in that the easier the break-in the lower the level of risk. It is in this regard that the level of security provided by locking devices is apparent. Some locks are simply more secure than others and present a more formidable barrier to surreptitious violations and physical assaults.

The Key-In-the-Knob

One of the most frequently used door locks today is the key-in-the-knob, though it is almost always the least secure type of available door locks. The typical key-in-the-knob lock is too susceptible to failure or breakage when force is applied to the knob or to the lock cylinder (figure 6-10).

Figure 6-10
Key-In-The-Knob.

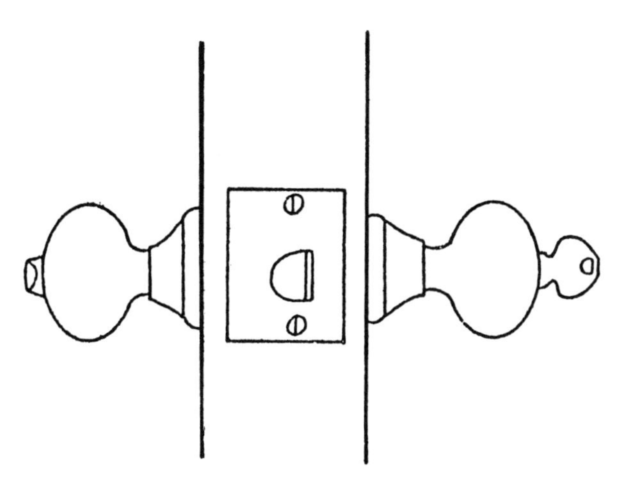

Auxiliary Locks

Auxiliary locks are added to a door or other opening and the existing locking device is left in place. Adding an auxiliary lock is usually the simplest way to bolster the security offered by the present, primary door lock. Two types of auxiliary locks are used most often.

A rim lock with a vertical or horizontal bolt-throw is a common auxiliary lock. When properly installed, a rim lock of the single cylinder or double cylinder variety with a deadbolt provides a substantial level of security. It is resistant to jimmying, prying, carding and jacking.

Another type of lock which is commonly utilized as an auxiliary lock is the tubular lock. The deadbolt variety, single or double cylinder, can be added with minimal difficulty and affords an improved degree of security over many primary lock types.

The Mortise Lock

To install a mortise lock, a cavity must be made in the door to receive the lock. The bolt is made so that a hole must be cut in the edge of the door. Mortise locks provide an acceptable level of security because most of the lock mechanism is in a metal enclosure in the door. They can be of the single or double cylinder variety and very often have additional functions such as automatic locking capability from one or both sides of the door (figure 6-11).

Figure 6-11
A Standard Mortise Lock.

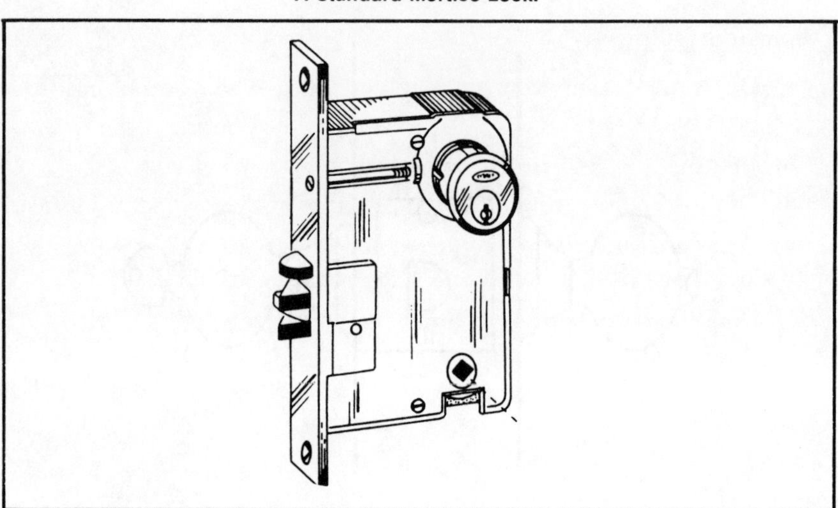

Window Locks

There are a number of specialty locks available for windows. Most windows when installed are provided with a latch, but these offer little resistance to pressure and most window latches can be forced open from the outside with a minimal level of skill. There are both keyed and non-keyed devices that afford a greater level of security than the latch. These devices are particularly applicable to ground floor windows or those near accessible stairs, fire escapes, etc.

The sliding glass door found in many homes and offices is both a window and a door. Generally the installation of these units and the manufacturer-supplied locking mechanism do not afford a high level of security. There are a variety of locking devices available for complementing the primary lock, including keyed and non-keyed devices, bars, and pinning devices. These will make it more difficult to raise, slide, or force the unit open.

Summary

Locks constitute a basic element for the security of any property or premises. Because there are so many types, brands, and models of locks available, one must be selective in the choice of a lock type. The security afforded by the lock, its cost and its intended use must first be determined before a good selection can be made. Whatever the locks and the keying system employed, they are only as adequate as those who use and control them.

Discussion Questions

1. Differentiate between pin tumbler locks, wafer locks, and lever tumbler locks.
2. Describe the elements of a master key system.
3. Outline and describe the responsibilities of the key control officer.
4. What are some of the important considerations when combination locks such as those found on safes or vaults are used?
5. Describe the workings of a pin tumbler lock.

Chapter 7
Electronic Alarm Systems*

Electronic protection of property and personnel has received increasing attention as a means to address a myriad of security problems. Electronic systems application to security has evolved to include almost every aspect of protecting people and property. Many such applications have multifaceted capability and are designed to integrate and interface a variety of sensory and surveillance components to protect the environment and its inhabitants.

Yet, even with the advances in electronic systems and devices, physical security and protection of people and property is still a matter of degree. Given enough knowledge, equipment, and time the criminal can still achieve almost any objective. Many times the electronic system chosen for a given environment and/or task fails to provide the expected level of security; the burglar is not detected, the robbery alert systems fails to function, or the access control system is easily compromised. The selection of a proper electronic system is not a simple task, particularly if the purchaser is not a security professional and/or is unfamiliar with electronic security systems and their applications to given environments and problems. The complexity of today's electronic devices, systems, and applications can be very confusing as the average user lacks the technical background and knowledge to make an educated purchase. Thus, there is great dependence placed on the security expert and/or the representative of an electronic systems company to recommend the proper system.

Any electronic system, whether it be for intrusion, detection, surveillance, environmental monitoring, or emergency alert, must be balanced against the nature and value of the property to be protected, the cost of the system, the presence and/or availability of police protection or security personnel, the integration of the system with the total environment, and the desired level of security. An evaluation of the total security program and its objectives must be accomplished so that the three major component areas of security, i.e., personnel services, security equipment, and electronic security systems will be balanced for maximal utilization.

---

* See end of chapter for a glossary of Alarm Systems and Devices Terminology.

Overdependence in one area may increase costs far beyond benefits, whereas underdependence may result in the creation of a "weak link" in the total security system.

Any protection system, whether personnel services, equipment, electronic, or some combination thereof, must satisfy two practical tests in any organization. The system must (1) work without major problems the way that it was planned and implemented, and (2) be economically feasible for the assets or facility that is being protected. Any protection system that does not work properly or is not economically feasible should be evaluated for improvement or elimination.

Electronic Alarm Systems (Functions and Selection)

In general, electronic devices are used to detect, monitor, or react to abnormal or predetermined security or environmental conditions in a facility. The function of an electronic alarm system can be viewed as one or more of the following: (1) detection of fire, (2) detection of intrusion, (3) emergency notification, and (4) monitoring of equipment or environmental conditions of the facility. Any one system may or may not incorporate all of the above. When utilized for any of these functions the elctronic system is operating as a machine and is performing a mechanized task. While various system components or devices may be better suited to certain protection tasks than man and their use may be an improvement in protection at less cost, the role of man in the total protection program cannot be eliminated. The human factor must remain for the capacity and capability of intelligent judgment to react and respond to environmental exceptions and problems signalled by the electronic system.

Thus, no inference should be drawn that electronics or man is superior one over the other. Obviously, routine, rote-like tasks should be relegated to electronic devices whenever possible, and activities requiring intelligence and judgment should be supplied by security personnel. The requirements, in terms of electronics, equipment, or personnel, of a particular security program can best be determined after such factors as the following have been considered:

- The threat environment, man-made or natural – what must the system protect against?
- The type of security needed, personnel, perimeter, access control, information – what is to be protected?
- The methods of security to accomplish given objectives or levels of security – what works best?
- The methods of coverage and response – what is to be the relationship of security equipment and electronics to security personnel?

- The resources available for security – what are the short-term and long-term costs?

Planning, then, is an essential and important element of the decision-making process of selecting the best combination and arrangement of people, equipment, and electronics for accomplishing protection and security objectives. The successful operation of any system depends on the proper integration, arrangements and relationships of its parts.

Types of Alarm Systems

Regardless of the operational features that an electronic alarm system may have, it will only be effective if there is a response to any signal initiated. Therefore, the termination of the electronic alarm signal, which indicates system reaction to an abnormal or problem condition, must be planned so that the proper personnel are alerted and a response is made. There are four basic types of terminating alarm system signals: local alarm system, central-station alarm system, proprietary central control system, and auxiliary alarm system.

The purpose, orientation, cost, and operational features of the electronic system will generally determine which of the four basic types of signal termination should be utilized. For example, if the primary purpose of a system was to detect fire, it would be foolhardy to install only a local alarm system in a remote, isolated area where no one would hear the signal (siren, horn, bell) and respond to the fire, whereas, a local alarm system designed to detect an intrusion to a facility having immediate response capability would frighten most intruders away. To be efficient and effective, any alarm system must be designed to meet the needs of the environment and the purposes for which it was intended. Decisions must be made regarding the orientation of the system, e.g., should it be oriented to catching the intruder via utilizing a silent alarm signal, would it be best to just scare the intruder away, or should the system have both signal capabilities, local and silent signal termination.

Local Alarm System

A local alarm termination system generally terminates on the premises at either a central control station or near the vicinity of the activated sensor. The sensor activates either a visual or audible signal or both. This system would require that someone be present at the facility at all times if an immediate response is necessary. Local alarms without monitoring personnel at or near the facility are usually undesirable for the following reasons:

1) Local alarms are generally very simple in design and structure. This allows even the amateur to easily defeat the system.

2) Audible alarms usually will not deter a person if that person knows the facility and location.
3) Audible alarms only act as a fear mechanism for persons who are surprised by the alarm. Even when a person is surprised by the audible alarm, intruders are rarely apprehended as a result of the alarm.
4) The problem of false alarms. Frequent false alarms do not build any good will between the company and the police or the neighbors.

If an outside audible alarm is used by a company without plans for a direct response to the alarm on the part of a company employee or an agent, i.e., security officer, it should be set on a timer so that the alarm will shut off after ten to fifteen minutes.

Central-Station Alarm System

A central-station alarm system is composed of fire, intrusion and/or monitoring sensing devices capable of activating a telephone, receiving module or screen at a location away from the facility that the system is protecting. Several alarm systems are generally monitored at a central location by trained personnel. The monitoring function generally requires that personnel observe the various alarm indicators twenty-four hours per day, seven days per week.

Most central-station alarm companies are privately owned, one of the largest being the American District Telegraph Company (ADT). Upon receiving an alarm from an activated sensor, the person monitoring the alarm panel will notify either the fire or police department, dispatch a security officer or do both.

Figure 7-1
Signal and Information Transmission in a Central Control Termination Arrangement.

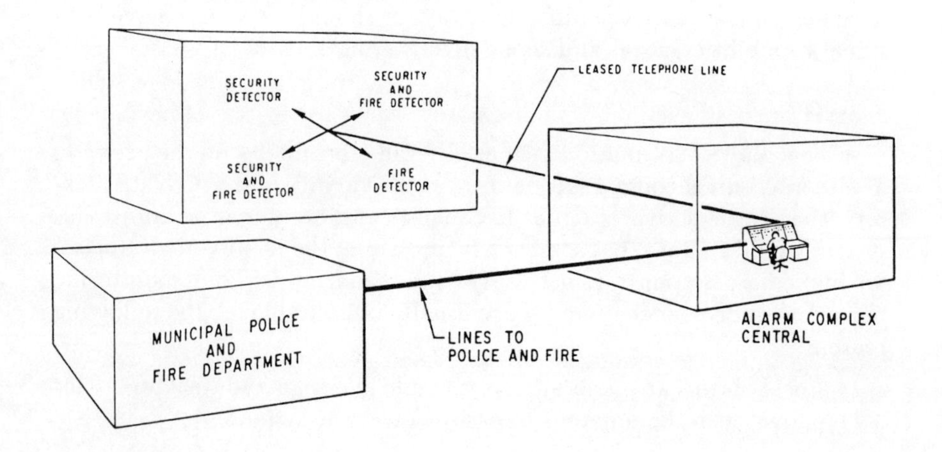

An activated fire sensor requires that the fire department be notified. If the activated sensor is an intrusion alarm, security personnel should proceed to the facility and wait on the outside of the building for police personnel before entering, if there is the possibility that an intruder may be inside. The security officer who has been dispatched by the monitoring person from the central station can alleviate the need for police response in cases where an employee set off the system or a malfunction has occurred in the equipment.

Proprietary Central Control System

A proprietary central control alarm system is one that is privately owned and controlled. This system is very similar to a central station system in that the monitoring of the alarm system is done at a central location, but the station is within the facility and not at a distant location. A central control center is usually staffed twenty-four hours a day, seven days a week and the response to any alarm is handled by dispatching facility personnel to the location of the alarm. These are in-house systems that serve only the owner, but they may in some cases protect more than one facility owned by the company.

In most cases, the response time to the activated alarm is less for a proprietary central control center alarm than with a central station alarm system. However, for all but the largest facilities, the proprietary system will cost more to operate than will the contractual service of a central station system.

Auxiliary Alarm Systems

Auxiliary alarm systems include all means of signal termination to police departments, fire departments or other designated locations. These methods include the programmed tape dialer, the digital dialer, and the "dry line" direct connect.

Auxiliary alarm systems involve notification directly to either the police station, fire department or another telephone number, and are usually silent alarm systems. Direct notification systems necessitate either line or wireless communications between the alarm system and the notification point. The usual transmission circuit is a telephone line leased for that specific purpose.

Tape dialer. The tape dialer alarm system is designed to dial a programmed telephone number when the alarm system is activated. The activation of the sensory device via the control unit releases the dialer and the programmed telephone number is dialed. Once the telephone number has been answered, a recording tells the receiver the coded message, i.e., the type and location of the alarm. The recorded message will repeat itself for a given number of times if the contact is not broken. A broken, cut, or malfunctioning telephone circuit usually will cause the system to fail.

Digital dialer. Instead of sending a programmed message, this unit transmits a coded message to a special receiver that can be located at the police department, fire department or other location. The coded message, usually numerical, indicates the location and type of alarm so that appropriate response can be made.

Direct connect. Direct connect alarm systems employ a telephone transmission circuit called a "dry line," which is an exclusive circuit connecting the alarm system directly to a specific location. The alarm signal is received by a module alarm unit, usually capable of indicating alarm activation or circuit malfunction.

While very common, auxiliary alarm systems have created problems for public services such as police and fire protection. Police and fire departments generally do not have the space for numerous direct connect module units, nor do they have the time or manpower to deal with the high rate of false alarms. Many police departments now refuse to monitor private alarm systems, and most that do require that alarm systems meet certain qualifications for quality, installation and service before they will monitor the unit.

Fundamentals of Alarm Systems

Alarm signal information, other than for the local system, generally is transmitted via leased telephone line to designated location(s). The regular telephone system in a facility may be used, or lines specifically dedicated to transmission of alarm system information. Often, these dedicated lines can be adapted to handle the two-way transmission of special, preprogrammed, supervised signals for the purpose of monitoring the integrity of the alarm circuit. The ability to monitor the status of the transmission line and its condition is essential to effective alarm system operation.

Another, less common, means of transmitting alarm signal information is via radio waves. Radio telemetry, transmitting by radio to a distant station and recording/receiving measurable, meaningful data and information, commonly referred to as wireless communication, is an alternative to telephone lines; however, the protected facility must have appropriate equipment which is generally quite expensive.

The internal features of alarm systems are composed of five basic components: (1) sensory device(s) that monitor and react to a change in the environment; (2) a control unit which acts as a signal processing unit; (3) an annunciator, either silent or local, that elicits human response; (4) a power supply from a commercial power source and/or alternative battery power source; and (5) circuitry, either hard wire or wireless, for transmission of signals. These five basic components are common to most alarm systems regardless of their function or purpose. Whether an alarm system is designed for detection of fire, detection of intrusion, emergency notification, or monitoring of equipment or facility conditions, the operating principles of each are much the same. It is possible to construct a very simple alarm system without all of these components, but in the vast majority of situations it would be impractical to do so.

Sensory Devices

Sensory devices initiate alarm signals as a result of sensing the stimulus or condition to which they are designed to react. When this sensing of a change in the environment or situation has occurred, a change in the flow of electrical current takes place. This is ordinarily referred to as completing or breaking the circuit.

Control Unit

The control units is the terminating point for all sensors and switches in the alarm system. It can be designed to have a variety of capabilities from a simple on-off switch to a complicated set of sensors and switches divided into zones and functions. Control units are usually housed in heavy steel, tamper-resistant containers. Elements of the control unit are arranged to receive signals from the sensors and to relay signals to the appropriate termination point. System control is accomplished by an exterior shunt lock to allow entry to the control unit to turn the system on or off, or by an entry-exit time delay component which allows time enough to get to the control unit to turn it on or off.

Annunciator

The annunciator is a visual and/or audible signalling device which indicates activation of the alarm system. Selection of the appropriate annunciation for an alarm system depends upon two factors: (1) the circumstances and location surrounding the alarm system site, and (2) the desired or required orientation of the alarm system. Very often the location of a facility or the availability of alarm services mandate that either a local alarm or remote, silent alarm be employed. When choosing the type of annunciation, particularly with an intrusion detection system, one must make the choice between an apprehension-oriented system (the silent alarm) and the deterrent-oriented system (the local alarm).

Power Source

The primary power source for alarm systems is obtained from commercial power sources. The 110 volt alternating current is transformed, rectified and filtered to provide direct current of the proper voltage to the alarm system. Thus, alarm systems are totally dependent upon an electrical power source for proper operation. In the event that the commercial power source is disrupted, an adequate standby power, either dry cell or rechargeable storage batteries, is an essential element of any alarm system.

Alarm Circuits

The alarm circuit transmits signals from the sensors to the control unit, which in turn transmits signals to the local or remote annunciator-receiving unit. Alarm systems are wired as either "open" or "closed" circuits. The open circuit system is a line without a flow of current present until a switch or relay is closed to complete the circuit. The closed circuit system is a line with current flowing through it and any change in this flow may initiate an alarm signal. Alarm circuits can also be installed by two different methods of circuit arrangement, i.e., the direct circuit connect and the McCulloh Loop. The direct method of wiring an alarm system is to connect each sensor or system to the control or receiving unit by an exclusive circuit. The McCulloh Loop is a circuit which has two or more sensors, switches, or systems on the same circuit.

Selection of Alarm Systems

Each type of alarm detection or notification system is intended to meet a specific type of problem. The necessity and feasibility of any alarm system must first be determined before installation begins. The following elements need to be considered:

1) Importance of the facility, materials, and processes.
2) Vulnerability of the facility, materials, and processes.
3) Appropriateness and feasibility of using specific types of alarm systems.
4) Initial and recurring costs of the alarm system compared to cost (in money or security) of possible loss of materials or information.
5) Savings in manpower and money over a period of time.
6) Response time by security personnel or other respondents.
7) Improvement over current security methods.

Decisions to utilize electronic alarm systems should lead to economy and improvement over existing security practices and methods.

Alarm Devices and Sensors

Alarm devices and equipment are usually classified in three general categories, according to the type of physical protection coverage provided: (1) point or spot protection, (2) area or space protection, and (3) perimeter protection. An electronic security system can provide the desired type and depth of protection by a combination of two or more of these categories.

The following discussion of alarm devices and sensors will be directed to the coverage, purposes and the functions that the alarm system is to provide and the types of devices, switches and sensors that will satisfy those objectives.

Intrusion detection alarm systems provide deterrence against and detection of unauthorized entry into a facility, building, or other structure. The situations and conditions at a particular site to be protected determine which devices or equipment would be efficient and practical. Following are some of the more common devices and the principles upon which they operate.

Electromechanical Devices (Perimeter-Point)

The most commonly used alarm sensors are electromechanical, i.e., sensors which operate on the principle of either breaking or closing an electrical circuit. Generally, the system operates in a way that requires a current-carrying conductor to be placed in a position between a potential intruder and the place to be protected.

Switches are commonly utilized in electromechanical alarm systems. The two most common types of switches are the simple contact switch and the magnetic contact switch. The simple contact switch requires only that the two halves of the switch device touch in order to complete a circuit. Any action to disconnect the switch will activate the alarm. The standard type of magnetic switch consists of a magnetically-activated switch unit and a magnet. The magnet is usually attached to a movable fixture and the "switch" unit to a permanent fixture. When moved, the magnetic portion of the unit will cause the switch component to either make or break the electrical circuit connection.

Another common electromechanical device is metallic window foil which can be applied to the glass in windows or doors. The foil tape cemented on the glass forms a complete circuit and if broken will activate the alarm system. A basic advantage afforded by window foil is the psychological deterrence it has on would-be intruders as it is obvious to outsiders that an alarm system is present.

Electromechanical devices are used primarily to provide perimeter protection for buildings or other structures. Individually, the devices provide point protection of a window, door, or other opening, but taken together they create a perimeter line of alarm protection.

Advantages of electromechanical sensor devices are:

1) Once installed, they provide relatively maintenance-free service. Environmental conditions will affect switches and exposed wire on outside units.
2) Not being sophisticated systems allows them to operate without excessive numbers of nuisance alarms.
3) They provide good perimeter security in low risk situations.

Disadvantages of electromechanical sensor devices are:

1) System can be easily compromised; walls and roof not usually covered.
2) Costly to install in facility with many coverage points.
3) Lack of local standards on installation and maintenance of various systems. A highly competitive business which may mean less quality.
4) Will not detect "stay-behinds" until they leave premises.

Photoelectric Devices (Space-Point-Perimeter)

The photoelectric type of intrusion detection device uses a light sensitive cell and a projected light source. The light beam is projected from a transmitter unit to a receiving unit which houses a photoelectric cell (figure 7-2). When an intruder crosses the beam, the contact with the

Figure 7-2
Standard Photoelectric Transmitter and Receiver Unit.

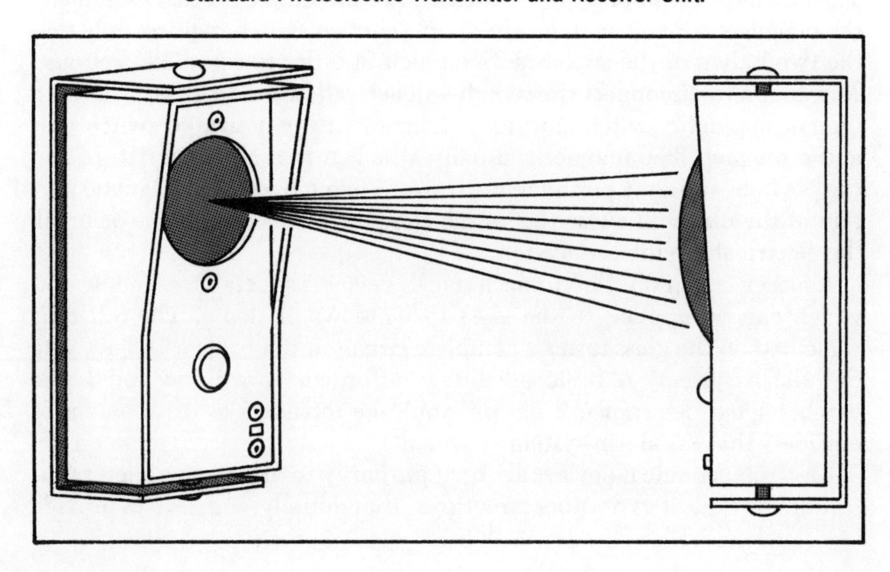

photoelectric cell is broken and the alarm is activated. The transmitter and receiver can be arranged in such a way that reflectors (mirrors) can be used to obtain a crisscrossed pattern of coverage. The light source can be white, infrared or laser. The most common one in use is the infrared because of its invisibility and ease of purchase. Coverage from infrared units can include point, perimeter or area depending upon their arrangement.

Advantages of photoelectric detectors are:

1) Useful at entrances, exits, and driveways where obstructive devices cannot be used.
2) When properly installed and used, can provide reliable security.
3) May be used to activate other security and/or safety devices, i.e, cameras and fire extinguishers.

Disadvantages of photoelectric detectors are:

1) When used outdoors, rain, fog, dust, and smoke can interfere with light beam.
2) Must be used in locations where it will not be possible to go over or under the light beam.
3) Requires frequent maintenance inspections of units and the grounds.

Audio Devices (Space or Point)

Sensitive microphones are installed in the protected area and are adjusted to tolerate the ambient sound levels in the environment. Attempts to force entry into the area generate sounds and noises that actuate the alarm. Audio devices can be used as audio monitors which provide the capability of "listening in" on the environment being protected. Contact vibration detectors are also included as audio devices since they detect vibration of sound in the structure being protected. Audio devices are particularly suited to protection of structures which are solid-walled and reasonably insulated from exterior noise.

Motion Detectors (Space)

The protection of an enclosed space can often be effectively accomplished by use of a class of alarm protection devices referred to as space alarms. Such systems derive their operating principles from a phenomenon known as the "Doppler Effect." There are two primary types of space alarms: the ultrasonic motion detector and the microwave motion detector.

The ultrasonic motion detector generates high frequency sound waves which fill a given enclosed area with a pattern of waves. Any motion within the protected area will compress or expand the transmitted sound waves, causing the reflected waves to differ in frequency from the original transmission. This change in frequency is detected and the alarm signal is activated.

The principles of operation for the microwave motion detector parallel those of the ultrasonic unit with one important exception. Microwave units utilize radio waves rather than sound waves, and are highly penetrating and not easily confined within many closed spaces. Buildings or rooms constructed of light materials would not be suitable to the utilization of a microwave motion detector.

Pressure/Stress Devices (Point)

Pressure sensitive devices are usually placed in a location where an intruder is likely to walk, i.e., in front of a door, under a window, in a hallway, or on the steps of a stairway. Usually these devices have the appearance of rubber mats in which are alarm circuit wires. When sufficient weight (about 40 pounds) comes to bear on the mat, circuit wires come into contact and an alarm signal is activated. Pressure mats can be disguised and used on bare floor, or they can be installed beneath carpet or rugs, thereby being invisible to intruders.

Other forms of pressure sensitive or stress sensitive devices can be used to detect changes or shifts in weight. Contact stress or vibration detectors can be placed under stairways or on floor joists to detect weight or stress changes. Pressure devices are also applicable to placement under art objects, store merchandise, etc., to detect removal, and on window frames, doors, safes, walls, etc., to react to vibration or shock.

Capacity Devices (Point)

The electomagnetic or capacitance type device can be installed on a metal fence, safe, file cabinet, or other metallic object. The protected metal object acts as part of the capacitance of a tune circuit. If a change occurs in the proximity of the protected object, such as the approach of a person, there is sufficient change in the capacitance to upset the balance of the system and cause an alarm. Unlike space alarms, the protective field around the protected object can be adjusted down to a depth of a few inches from its surface. In this way objects can be protected day or night, even during business hours. Capacity devices have a high degree of security but are restricted in application because they can only be applied to ungrounded metal objects.

Glass Breakage Sensors (Point or Perimeter)

Glass breakage sensors are shock/sound sensing devices which are attached to the glass and sense the breaking of glass by vibration or frequency of sound. The sensor mechanically responds to the breaking of glass and can be used to protect perimeter windows, glass storage units, etc.

Passive Infrared (Space)

Passive infrared units monitor an area by detecting a change in the normal radiation/temperature environment of the area being protected. The unit does not transmit a signal but is a receiving device capable of detecting relative changes in radiation temperature, such as a human body moving through the field of coverage. Passive infrared units can be used to cover large open areas, hallways, offices, etc.

Vibration Detectors (Point)

Vibration detectors are normally mounted on fixed, stable surfaces. When the sensitivity of the vibration device is properly adjusted, any shock to the surface beyond the set tolerance level will cause the contacts to touch or separate, signaling an alarm condition.

To discuss all possible types and varieties of intrusion detection devices is beyond the scope of this text. However, those presented represent the devices that are most commonly utilized to detect or deter unwanted intrusion, and a basic understanding of them is essential to the selection and application of an appropriate electronic intrusion alarm system.

Fire Detection Systems

Fire is one of the most destructive forces faced by man. Whether its initiation is man-made or natural, potential losses of life and property can be enormous. Of course, steps to prevent the occurrence of fire precede those steps taken to detect or suppress fire. However, the detection of fire, if and when it occurs, is a necessary step that must be taken in homes, business, and industry.

Alarm systems and devices designed to detect fire may be incorporated into or parallel intrusion alarm systems. Methods and types of signal transmission and system components are generally the same. The exceptions are the respondents to fire alarm signals and the detection devices employed.

The fire department or a fire brigade are the ultimate respondents to a fire. Local alarms should always be included in a fire detection system even if a central station, proprietary central control station, or auxiliary system is also used. With fire detection systems the objective is not apprehension, but the fastest possible response.

Figure 7-3
SUMMARY CHART
Detectors

Device	Application	Operation	False Alarm Potential	Advantages	Disadvantages
Contact Switches	Primarily openings: doors, windows, etc.	Metallic Contact held in closed or open position by magnet component of two-part device. Mechanical contacts operate in similar manner except a spring, pressure release mechanism, etc., causes the switch to function.	Contact switches are basically reliable environmental conditions or excessive or little to no operation may cause unit to malfunction or deteriorate.	•Cost is comparatively low •Flexible application of usage •Dependable and maintenance free operation •Visibility of devices enhances psychological deterrence	•Subject to compromise and circumvention •Excessive or little usage may cause operational problems •Surface mounted devices may not be decorous
Shock Sensors	Primarily used on openings and storage units.	Contact mechanism that reacts to frequency and severity of shocks and vibrations.	Sensitivity adjustment is critical, particularly if environmental conditions vary. Basically reliable operation.	•Relatively easy to install •Flexible in terms of application to surfaces •Variable cost, depending on sophistication of device	•Can be expensive •Sensitivity level may need frequent adjustment
Traps	Openings or passageways such as duct work, false ceiling areas, window-mounted air conditioning units, storm drains, etc.	A wire or cord under tension that when cut, loosened, or tightened, will activate the alarm circuit.	Reliable Potential usage areas could be subject to false alarms caused by animals or environmental extremes.	•Can be constructed to protect points of potential intrusion not applicable to usage of other devices •Low cost	•Must be physically repaired or manually reset when activated
Foil	Normally mounted on glass; can be	Metallic foil-like tape which carries	Excessive vibration can cause deteriora-	•Low cost of materials	•Time consuming installation •Easily damaged

	applied to other smooth surfaces.	current. Breaking or cracking of tape activities alarm circuit.	tion of tape. Extreme environmental conditions reduce effective life expectancy.	•Easy to install •Visible psychological deterrent •Reliable under normal conditions	•Open to circumvention
Glass Breakage Sensors	Glass surfaces.	Responds to frequency of breaking glass. Acts as a contact microphone.	Relatively stable if properly adjusted. Does not need power source.	•Relatively low cost •Coverage of large glassed area possible •Easy to install	•Subject to false alarm in hostile environment
Audio Devices	Mounted on fixed, stable surface, usually the wall.	A microphone which responds to sound levels above set tolerance levels. Capable of discriminating between certain sound frequencies.	Reliable when tolerance sensitivity is properly adjusted.	•Volumetric coverage •Can be used to listen — into the environment being protected •Monitoring personnel can record sounds, conversations, etc., of "intruders"	•Relatively expensive •False alarms due to infrequent or unusual environmental noises
Vibration Contacts	Mounted on surfaces subject to forced entry; doors, window frames, walls, ceilings, safes, vaults, etc.	Excessive vibration causes metallic contacts to separate and activate the alarm circuit.	Sensitivity adjustment is critical.	•Relatively low cost •Fairly reliable if properly adjusted •Easy to install	•Environmental conditions are often too extreme or variable for usage.
Ultrasonics (Motion Detector)	Unit generally mounted on stable surface, positioned to focus on area to be protected.	Transreceiver unit detects change in frequency of transmitted sound waves. Change in frequency difference caused by intruding object (Doppler Principle) activates alarm circuit.	Overly sensitive if improperly adjusted. Environmental conditions such as air movements, animals, birds, etc., can cause false alarms.	•Volumetric coverage •Non-penetrating waves •Difficult to defeat	•Relatively expensive •Coverage of dead spots •Sound waves may be overly absorbed by some materials

Microwave (Motion Detector)	Unit generally mounted on stable surface in fixed position, and positioned to focus on area to be protected.	Transmits and receives radio waves. Change in frequency difference of transmitted waves caused by intruding object (Doppler Principle) activates alarm circuit.	Penetrating and reflective of radio waves outside the area of intended coverage. Cause unwanted alarm condition. Sensitivity adjustment is critical.	•Volumetric coverage •Difficult to defeat	•Relatively expensive •Penetrating waves •Reflected waves may bounce out of coverage area
Passive Infrared	Normally mounted on fixed, stable surface and positioned to cover area to be protected	Monitors area of coverage for a relevant change in temperature/infrared energy.	Stable in proper environment.	•Volumetric coverage •Infrared energy will not penetrate most materials	•Coverage area must be line of sight •Reacts to animals and hot/cold air movements
Photo-electric (Infrared)	Transmitter and receiver units mounted on fixed, stable surface. Can be arranged to provide point, perimeter, or area coverage.	Pulsed infrared light is transmitted to the receiver unit which contains a photocell. Interruption of light beam causes photocell to react and signal alarm condition.	Stable when alignment is maintained.	•Very versatile in application •Relatively inexpensive •Use of reflectors enables unit to provide crisscross pattern of coverage, even around corners	•Coverage area must be line of sight •Reacts to animals or objects crossing light beam
Pressure Mats	Normally placed in doorways, under windows, on stairways, etc. May be placed under carpet or rugs.	Pressure applied to mat causes two metal strips (contacts) to touch and activate alarm signal.	Stable.	•Inexpensive and easy to install •Can be disguised or hidden •Versatile in protection of perimeter openings, safes, vaults, and traffic corridors	•Subject to rapid deterioration in heavy traffic area •Will short out when wet •Restricts placement of furniture and materials

Fire Detection Devices

Heat detectors are of two general types: those that operate at a predetermined temperature, called fixed temperature devices, and those that operate when there is an unusual increase in temperature, designated as rate-of-rise types. The fixed temperature detector used most widely in protective signaling systems employs the principle of different coefficients of expansion in metals, similar to a common thermostat. By arranging electrical contacts so that a circuit opens or closes according to the difference in length of two strips of different metals, an alarm signal will occur at a predetermined temperature. Other types depend upon heat expanding a metal disc until electrical contact is made, or on the melting of heat-sensitive plastic insulation.

Rate-of-rise detectors react to a sudden, rapid change in temperature and can be set to operate more readily than fixed temperature detectors. There are two principal types of rate-of-rise heat detectors. The first consists of a pneumatic tube which reacts to changes in air pressure caused by heat, and the second uses the principle of the thermocouple.

Smoke Detectors

In many fire situations detection can be more rapid if smoke detectors are present. The two most common types of smoke detectors are the photoelectric unit and the ionization unit. The photoelectric smoke detector is activated by smoke passing through a photoelectric beam. Sufficient concentration of smoke will interrupt the beam and cause activation of the alarm. The ionization smoke detector reacts to hydrocarbons that develop in the chemical processes prior to actual ignition.

Other fire detection devices include the rate compensated detector, laser beam fire detector and ultraviolet or infrared flame detector.

Emergency Notification Systems

Instead of being triggered by an intruder or fire, an emergency notification system is used for security and safety in unusual or dangerous circumstances. Perhaps the most common example of this is the robbery alert system used by businesses and particular financial institutions.

A robbery alert system is usually a component of the intrusion detection alarm system. For this to be possible the master control unit must have a day and night circuit, which allows zoning of alarm devices and alarm circuitry. Those intrusion detection devices which would interfere with daytime business activities are on the night circuit and those which are to be utilized for the robbery alert system are on the day circuit. By setting the control unit on the day circuit and by strategic placement of hold-up buttons or other devices around the facility one can alert the police or others to a robbery in progress. Robbery alert signals should be silent and should only be used if it is safe to do so (figures 7-4 and 7-5).

Figure 7-4
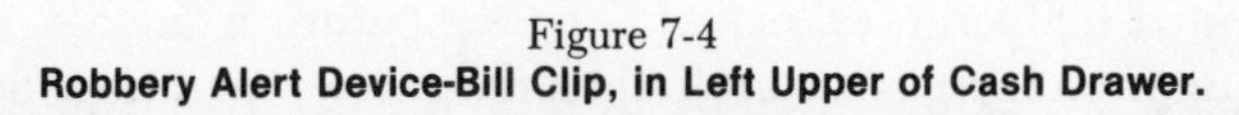
Robbery Alert Device-Bill Clip, in Left Upper of Cash Drawer.

Another example of an emergency notification system is that utilized by business and industry to alert employees and others that an emergency has occurred or is pending. Alarms must be audible or visible to everyone if effective action is to be taken. Steam whistles, bells, sirens and other devices are utilized to warn people to take cover, to evacuate, or to take some other action. Such systems and devices are used to warn of fires, explosions, tornadoes, hostile attack or any other situation with hazardous potential.

The Problem of False Alarms

Although alarm systems have been proven effective in deterring and apprehending burglars, they are subject to certain inherent problems. A traditional and still-to-be-resolved problem is that of false alarms. It is estimated that from 90 to 98 percent of all alarms transmitted are false. This high percentage can be attributed to three factors: (1) user error or negligence, (2) poor installation or servicing, and (3) faulty equipment.

More than half of all false alarms are estimated to result from user error or negligence. Users often do not understand how to properly operate their systems. Commonly, alarms are set off by users who fail to lock doors or windows or who enter a secured area when the system is engaged. Merchants have been known to use their alarm systems to summon the police to deal with bad checks or suspicious individuals. Some users even set off their alarms to time police response to their premises.

The second factor contributing to false alarms is poor installation or servicing. In order to function as intended, an alarm system must be properly installed and maintained. Equipment that is installed in an inappropriate environment or improperly positioned, set, or wired, is more likely to produce false alarms. Likewise, if equipment is not adequately maintained, the chances of false alarms increase. Too often, installers and service personnel lack the necessary skills and knowledge for today's more sophisticated equipment.

Figure 7-5
Robbery Alert Button Concealed Beneath Counter.

Poor installation or servicing has caused states and local governments to develop standards for installing and servicing alarm systems. In addition to standards for both installers and the installation procedures, some type of bond is now commonplace. Future standards in the alarm business will probably be as stringent as those for electricians and plumbers.

The third common cause of false alarms is faulty equipment. If equipment is electrically or mechanically defective, the alarm can be activated when, for instance, the equipment breaks or shorts out the circuit. The use of cheap, substandard equipment frequently leads to false alarms.

In addition to false alarms that can be traced to the above factors, there are a certain number of false alarms whose cause cannot be determined. Based on the results of various studies, roughly 25 percent of all false alarms fall into this unknown category. It is possible that they may, in fact, be the result of user error; faced with probable sanctions, a user may deny responsibility for a false alarm. Another possibility is that a burglary or other unauthorized intrusion may have been successfully prevented, leaving no visible evidence of intrusion or attempted entry.

The continued high incidence of false alarms, whatever the cause, has led to other problems. In the use of automatic telephone dialer alarm systems, a large number of storm-caused false alarms simultaneously occurring can tie up police trunklines and switchboards, seriously hampering police capacity to respond to genuine emergencies. Malfunctioning of such systems can lock-in police communications trunklines for considerable periods of time. Although telephone dialers offer effective, low-cost protection, these problems have created negative police reaction toward their use.

A more serious problem, especially to police personnel, is that of the personal risks involved in false alarms. The high-speed response to false alarms unnecessarily endangers the personal welfare of police, as well as other drivers and innocent bystanders. False alarms often also bring to the scene alarm company respondents who are frequently armed, presenting a further threat to personal life and safety.

Another problem of false alarms is the burden of expense to the users of alarm systems. When a system false alarms, servicing is usually required, resulting in increased costs that are eventually absorbed by the users. Further, some local governments directly impose fines upon users whose systems repeatedly produce false alarms.

Monitoring Systems

There are many processes and pieces of equipment that are critical to the continued operation of businesses and industrial concerns. Often these need to be monitored by an electronic system capable of sounding or sending an alarm signal. The necessity of monitoring may be due to economic reasons or to the hazardous nature of the process or operation. The importance of this particular type of electronic alarm detection/monitoring system is best illustrated by the nuclear power industry.

Particularly applicable to monitoring systems are closed circuit television and other types of surveillance equipment. CCTV would allow the remote monitoring of an environment, equipment or process without danger of exposure to personnel. In addition, CCTV is particularly applicable to retail operations as both a deterrent and detection device for shoplifting, and if provided with a video tape recorder (VTR) provides a visual record of the criminal act. Today, CCTV systems are in wide use in hospitals, motels-hotels, financial institutions, office buildings, industry and other areas. A CCTV system in basic form consists of a television camera, a monitor, connecting circuitry, and a power source. An expanded system with numerous cameras, monitors, VTR, remote control, etc., is expensive yet may offer improvement and savings over existing security methods (figures 7-6 and 7-7).

Figure 7-6
Security Control Center with CCTV Monitors.

Underwriters' Laboratories

Underwriters' Laboratories is an independent non-profit service corporation which applies existing safety and testing standards to products submitted by manufacturers. A UL label on a product usually indicates that its design, and the manufactured item itself, meet certain Underwriters' standards. Thus, in order to satisfy criteria for alarm system certification, protective devices, control units, circuitry, etc. must meet specific UL burglary prevention and detection standards. In addition to certification of alarm and detection devices, UL certifies local and central station alarm service companies. Alarm installing companies and service companies must utilize UL-approved products, install alarm systems according to UL standards, provide maintenance and inspection as required, and satisfy certain performance standards for alarm responses.

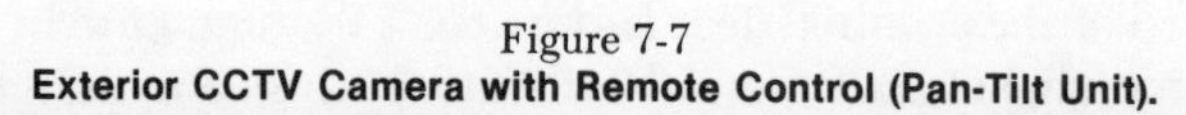

Figure 7-7
Exterior CCTV Camera with Remote Control (Pan-Tilt Unit).

Summary

An understanding of the applications, operations, and principles of electronic alarm systems is essential to effective and efficient usage. There are a variety of systems and devices available, and more are always being developed. All have weak points by which their functioning can be minimized or even completely disrupted, yet most will prove satisfactory if properly selected, installed, and maintained.

The use of electronic warning and/or detection systems has been proven beneficial in a wide variety of situations; however, individuals responsible for security planning and application must carefully analyze their security needs and the systems available before making a decision.

Discussion Questions

1. What are some of the key considerations when deciding whether to deploy a particular intrusion system?
2. Briefly describe the functions of alarms and alarm systems.
3. List and briefly describe the components of an alarm system.
4. What are some of the elements to be considered when selecting an alarm system?
5. Describe the workings and advantages of capacity devices.

Alarm Systems and Devices Terminology

Access Control – The control of pedestrian and vehicular traffic through entrances and exits via means of security personnel, equipment, and/or electronic means.

Active Sensor – A sensing device which detects a disturbance within the field of its transmitted signal.

Actuator – A switch that is manually or automatically activated to transmit a signal.

Alarm Circuit – An electrical circuit of an alarm system which transmits an alarm signal.

Alarm Condition – A sensory device or alarm component has reacted or been activated to signal a threatening condition.

Alarm Discrimination – The ability of some alarm systems and devices to distinguish between the ambient (normal) and abnormal conditions of the environment being protected.

Alarm Drop – Alarm system has activated and is in alarm condition.

Alarm Signal – A signal received and processed by the control unit to indicate an alarm condition.

Alarm Station – A device installed at a fixed location which can be manually activated to transmit an alarm signal.

Alarm System – An assemblage of components designed and arranged to monitor and signal an alarm condition.

Annunciator – The component of the alarm system that signals an alarm condition has been received. The annunciator signal may be audible, visible or both. Its primary purpose is to elicit human response to the alarm condition.

Area (Space) Protection – Protection of an area, other than perimeter or point. Sensory device is capable of protecting a large area or space.

Audible Alarm Device (Annunciator) – An annunciator device that makes noise, e.g., siren, horn, or bell.

Audio Detection System – An alarm system capable of monitoring and sensing sound in the area being protected.

Auxiliary Alarm System – Includes all means of alarm system signal termination to police departments, fire departments, or locations other than central stations, proprietary control centers, and local alarm systems. Alarm signals are generally transmitted by means of tape dialer, digital units, or dryline direct connect.

Bill Clip – A holdup alarm device, located in the cashdrawer, that is activated when bill(s) are removed allowing contact to be made.

Bug – A microphone or other audio sensor used for the purpose of "listening in" or audio surveillance.

Capacitance Alarm System – A system in which the object being protected is electrically connected to a capacitance sensor. Intrusion into the capacitance field around the object causes a change in the electrical field and an alarm condition (sometimes called a Proximity Alarm System).

Central Station – A control center to which subscriber alarm systems are connected and monitored. Central stations are owned and operated independently of the facility being protected. Subscribers pay a fee to the central station according to the level or amount of alarm services being provided.

Circumvention – A means of bypassing or defeating the alarm system.

Closed Circuit – An alarm circuit that has a continuous flow of electrical current.

Closed Circuit Television (CCTV) – A self contained system of electronic surveillance including camera, monitor, and circuitry. Additions to system may include pan-and-tilt to control coverage capability of camera, zoom lens on camera to obtain close-ups, etc.

Contact – A switch or relay that causes an interruption or completion of current flow by the touching or separation of its metallic parts.

Contact Microphone – A listening device (microphone) designed for attachment to the surface of the area to be protected.

Control Unit – The brains of the alarm system which functions to monitor, receive, transmit, and activate components to achieve system objectives.

Crossover – An insulated connecting path used to connect foil tape across dividers in multiple pane windows.

Detection Range – The outermost distance at which a sensor will effectively detect an intruder.

Digital Dialer – A device used to transmit alarm signals to a special unit capable of receiving coded messages that indicate the location and type of alarm condition.

Direct Connector – An exclusive telephone transmission line connecting the alarm system directly to a specific location.

Door Cord – A short, insulated cable with attaching block and terminals used to conduct current to a device which is mounted on the movable portion of a door or window.

Doppler Effect – A change or shift in transmitted and received frequencies caused by movement within the protected area. Primary to the operation of motion sensors such as the ultrasonic and the microwave.

Duress Alarm Device – A device that is normally manually activated and produces either a silent or local alarm. Generally used to indicate a condition of personal stress or emergency, such as fire, robbery, illness, etc. (Also called a "panic" alarm) The actuating device may be fixed or portable, wired or wireless.

Electromagnetic - Pertains to the characteristics and relationship between electrical current flow and magnetism.

Entry Time Delay - The amount of time between actuating a sensory device on an entrance and the transmission of an alarm signal by the control unit. The time delay allows a person with a control key to enter and turn off the control unit without causing an alarm.

Exit Time Delay - The time between turning the control unit on and exiting a protected door without causing an alarm signal. The delay is provided by a timer within the control unit that is set to accommodate the situation regarding both exiting and entering.

False Alarm - An alarm signal transmitted when no alarm condition exists. False alarms may be due to equipment failure or malfunction, environmental conditions or human error.

Foil Tape - Thin metallic strips affixed to a protected surface, usually glass, and connected to an electrical circuit. If the tape is broken, e.g., by a crack in the glass, an alarm signal is initiated.

Foot Rail - A common holdup alarm device used by banks at tellers windows. When confronted by a robber, the teller can activate the foot rail with foot pressure to initiate an alarm signal.

Glass Breakage Detector - Essentially a contact microphone attached to window glass to detect breakage of the glass.

Heat Sensor - A sensor which responds to a predetermined level of temperature or to a rate of temperature increase greater than a preselected rate of rise.

Holdup Alarm Device - A device, fixed or portable, which is used to signal a robbery. The device ordinarily initiates a silent alarm signal and may take the form of hidden buttons, footrails, hand-carried activators, bill clips, etc.

Infrared Detector - A passive sensory device which detects changes in radiation/temperature of the area being protected.

Interior Alarm Protection - An alarm system which provides protection of the interior of a building. Areas of coverage would include the interior perimeter, doors, file cabinets, safes, vaults, etc.

Intrusion - Any unauthorized or illegal entry into a protected area.

Ionization Smoke Detector - A sensory device which contains a small amount of radioactive material that reacts to smoke particles entering the ionization chamber. A sufficient amount of smoke in the chamber decreases the conductance of the ionized air and when a predetermined level of nonconductance is reached, the detector circuit is activated.

Lacing – A network of fine wire surrounding the area to be protected, such as a vault or safe. The network of wire forms a complete electrical circuit. Usually the wires are embedded or covered by concrete, paneling, plaster, etc., so that their presence is not obvious to a would-be intruder.

Line Supervision – The capability of monitoring an alarm circuit and/or transmission line to detect changes in the circuit characteristics.

Local Alarm System – An alarm system that produces a signal in the immediate area of that being protected. Such signals should be audible and/or visible for the purpose of eliciting human response to the alarm condition, and, hopefully, sufficient enough to "scare" the intruder away.

Magnetic Alarm System – An alarm system that is capable of detecting a change in a magnetic field when ferrous objects such as guns, knives, bombs, etc., enter the area of coverage. One such device, the magnetometer (metal detector) can be used and designed to search for buried coins and metal or in another form to screen airline passengers for hidden weapons.

Magnetic Switch – A switch which consists of two separate parts, a magnet and a magnetically operated switch. The switch component of the unit is normally mounted on the stationary portion of a door or window directly opposite the magnet, which is mounted on the movable segment of the opening. When the movable portion of the door or window is opened, the magnet moves away with it, causing the switch to make or break its contact.

Mat Switch – A pressure-activated switch normally used under carpeting or disguised as an entry mat. May be designed for small or large areas of coverage.

McCulloh Loop – An electrical circuit having several devices on the same "loop."

Mercury Switch – A switch containing mercury, which when vibrated or tilted makes or breaks the electrical contact. Mercury contact switches can be used on doors, windows, safes, fences, stairs, storage units, etc.

Microwave Motion Detector – An active sensory device which is capable of detecting via the "doppler principle." High frequency radio waves are projected into an area of coverage and any penetration of the protected area causes the transmitted signal to be disturbed, and subsequently detected by the unit.

Motion Sensor – A sensor which responds to motion.

Normally Closed Switch – A switch in which the contacts are in the closed position when the alarm system is "on" and in a secure condition.

Normally Open Switch – A switch in which the contacts are open (apart) when the alarm system is "on".

Passive Sensor – A sensory device that does not transmit a sensing mechanism, rather it monitors and reacts to changes in the environment, e.g., audio sensors, vibration sensors, and infrared sensors.

Perimeter Alarm System – An alarm system which provides coverage of the outermost boundary of the area being protected.

Photoelectric Alarm System – An active sensory device that uses a light beam (white, infrared, or laser) and photoelectric sensor to provide a specific line-of-sight protection. Beam of light may be constant or modulated, and area of protection may be increased by using a series of reflectors. An interruption of the focused beam by an intruder is sensed by the photoelectric sensor and an alarm signal is generated.

Photoelectric Smoke Detector – A beam of light is projected to a photoelectric cell within the detector unit. A sufficient amount of smoke will interrupt the light beam and cause actuation of the alarm system.

Point (Spot) Protection – Alarm system protection of specific objects such as doors, windows, safes, etc.

Proprietary Central Control – An alarm system similar to the Central Station except it is owned and operated by the facility being protected. One or more facilities may be protected by the same proprietary central control station.

Radar Alarm System – An alarm system or tracking device that utilizes radio waves to detect motion and/or speed of motion.

Remote Alarm – An alarm signal that is transmitted to a location removed from the area being protected.

Seismic Sensor – A sensory device generally buried in the ground which responds to vibrations within the earth. Will react to an intruder walking within the detection area, or if designed to do so, will aid in the detection and measurement of earthquakes.

Sensor – A device designed to respond to given events or stimuli.

Shunt – Generally a key-operated switch which isolates (turns off) a portion of the alarm system. Used most often for the purpose of entering or exiting protected doors.

Silent Alarm – An alarm signal that is transmitted to a remote location. The orientation of silent alarm signals is generally apprehension of the intruder. In some situations, such as a hold-up, the initiation of a local alarm would be dangerous for the victims.

Smoke Detector – A device capable of detecting visible or invisible particles and products of combustion.

Spring Contact – A contact device using a current-carrying spring which monitors the position of the object being protected, e.g., door, window, file drawer, etc.

Standby Power – Equipment capable of producing sufficient electrical power to operate a system in the event the primary power source fails. May be supplied by batteries, generators, or some combination of the two.

Stress (Strain) Detector – A sensor capable of reacting to applied stress such as the weight of an intruder walking across the floor or on a stairway.

Supervised Alarm System – A system which monitors conditions and procedures of system usage for variations or deviations from the norm.

Surveillance – Monitoring of premises through the use of various alarm system devices, particularly CCTV.

Tamper Device – Any switch or device constructed to detect any attempt to gain unauthorized access to the alarm system or manipulate its operation.

Tape Dialer – A device used to transmit alarm information via means of programmed telephone messages to designated locations. Activation of a sensory device and operation of the control unit releases the tape dialer to dial programmed telephone number(s) with a recorded message of the alarm condition and its location.

Trap Device – Usually a switch type device installed to protect openings or situations not conducive to other means of protection, e.g., a cord stretched across a window air conditioner unit with one end of the cord having a piece of nonconducive material between two metal contact points which in turn are a part of the alarm circuitry. An attempt to remove the air conditioner would pull the cord away and the nonconductive wedge from between the two contacts allowing for the activation of the alarm signal.

Trickle Charge – A continuous direct current sufficient enough to maintain a battery at full charge.

Trip Wire – A switch which is actuated by breaking, removal, or release of pressure of a wire or cord stretched across an area.

Ultrasonic Motion Detection – An active sensor which transmits and receives ultrasonic sound waves. The device operates by filling a space with a pattern of sound waves and reacting to changes in the pattern of the waves caused by a moving object.

Vibration Detection Sensors – Contact microphones on vibration sensors are attached to fixed surfaces and react to excessive levels of vibration caused by sound, attempts at forcible entry, and other vibration-causing forces. Such units must be adjusted to compensate for the normal conditions of the area being protected.

Volumetric Sensor – A sensor with an expanded detection zone, capable of protecting a large area.

Walk Test Light – A feature of most motion detectors which allows the user to set the sensitivity/coverage level of the unit. It also allows for routine checking of the units.

Zone – The area being protected by an alarm system can be divided into zones. Division allows monitoring personnel to better pinpoint the location of an alarm condition and zoning allows for flexibility in system usage.

Chapter 8
Security Storage Containers and Information Security

The final line of defense at any facility from theft, robbery or fire is the high security storage area. The degree of desired security for any facility will be dictated by the items or valuables to be protected. Even though every facility will have unique security requirements, there are general principles which apply to any security program.

The choice of type and quality of security storage container for a facility will depend on what is being protected. A security storage container for valuable paper documents will be different from one for diamonds or precious metals. A container which is fire resistant would be more appropriate for the paper document, whereas a burglary resistant unit would be required for the diamonds or precious metals.

It is important to recognize that security storage containers are generally either fire resistant or burglary resistant but not both. Each storage container usually provides only one specialized function and provides only a minimum amount of protection in the other area. If both fire and burglary protection are not deemed necessary, then costs can be greatly reduced by having a container which will provide only one type of protection.

Safes

Safes are very expensive and great attention must be given to the particular needs before an investment is made in a safe.

Safes are designated alphabetically in two categories to describe the degree of protection they provide: (1) fire resistive and (2) burglary and robbery resistive. The rating of safes in the area of fire protection is not mandated by any federal or state law or code, but through the Safe Manufacturers National Association (SMNA) specifications and independent tests conducted by Underwriters' Laboratories.

Fire Resistive Safes

The SMNA provides specifications for the manufacturing of fire resistant or resistive containers and Underwriters' Laboratories does the independent testing. A U.L. label or rating means that the merchandise in that class meets the minimum fire specifications designed for that class by SMNA. Table 8-1 contains listings for SMNA and U.L. labeled fire resistive safe equipment.

Table 8-1
Safe Manufacturers National Association
Fire-Resistive Labeled Equipment[1]

Product Classification	SMNA Spec.	SMNA Class.	U.L. Rating	Test Feature
Fire Resistive Safe	F 1-D	A	Class 350 - 4 hr.	Impact
Fire Resistive Safe	F 1-D	B	Class 350 - 2 hr.	Impact
Fire Resistive Safe	F 1-D	C	Class 350 - 1 hr.	Impact
Insulated Filing Device	F 2-ND	D	Class 350 - 1 hr.	No Impact
Insulated Filing Device	F 2-ND	E	Class 350 - ½ hr.	No Impact
Insulated Record Container (Ledger File)	F 1-D	C	Class 350 - 1 hr.	Impact
Insulated Record Container	F 2-D	C	Class 350 - 1 hr.	Impact
Insulated Record Container	F 2-D*	Class 150	Class 150 - 4 hr.	Impact
Insulated Record Container	F 2-D*	Class 150	Class 150 - 2 hr.	Impact
Insulated Record Container	F 2-D*	Class 150	Class 150 - 1 hr.	Impact
Fire Insulated Vault Door	F 3	6 hour	Class 350 - 6 hr.	—
Fire Insulated Vault Door	F 3	4 hour	Class 350 - 4 hr.	—
Fire Insulated Vault Door	F 3	2 hour	Class 350 - 2 hr.	—
Fire Insulated File Room Door	F 4	1 hour	Class 350 - 1 hr.	—

NOTE:

Class A	protects paper records from damage by fire (2,000°F) up to 4 hours.
Class B	protects paper records from damage by fire (1,850°F) up to 2 hours.
Class C & D	protects paper records from damage by fire (1,700°F) up to 1 hour.
Class E	protects paper records from damage by fire (1,550°F) up to ½ hour.
Class 150	protects EDP records from damage by fire and humidity for rated period.
The Drop or Impact test:	The Drop or Impact test is used to determine whether or not the fire resistance of a product would be impaired by being dropped 30 feet while still hot. Fire resistant equipment is designed specifically to resist fire, and consists of a metal insulation.

*Impact tested unloaded.

Class "A", fire resistant safe, SMNA specification F 1-D, U.L. rating, class 350-4 hours, and with a test feature of "Impact" means that this safe would withstand temperatures of up to 2000°F. for a period of four hours. During the span of four hours (0-4 hours) the temperature of the interior of the safe would not exceed 350°F. Since 350°F is the ignition point of paper, any record or paper content would be considered to be protected under the above conditions. This same safe would also withstand an impact of being dropped 30 feet while still hot. This feature would be very important in a multi-storied building where fire damages the floor and allows the safe to fall through. A Class "A" or 350-4 hour fire resistant safe provides the maximum fire protection available in a safe. It is recommended that this safe be used when the following conditions exist:

1) Facility is located in a remote area where response time will be lengthy and no firefighting personnel are available in the premises.
2) Facility has extremely valuable papers or records which either cannot be replaced or can be replaced only after costly delays.
3) Facility has materials, substances or petroleum products which will cause a quick increase in temperature during a fire.
4) Facility has any unique features which would make it difficult to easily and quickly extinguish a fire, e.g., a grain elevator building.
5) Facility has multiple floors and the safe is stored on an upper level.

Class "D", insulated filing devices, SMNA specification F 2-ND, U.L. rating, Class 350-1 hour, and with a test feature of "No Impact," will withstand temperatures of up to 1700°F for a period of one hour. During the span of one hour the temperature of the interior of the safe would not exceed 350°F. This safe would *not* withstand an impact or drop. Class "C" is the most popular and commonly used safe.

Fire resistive safes are double-walled containers. Between the outer and inner metal walls is a layer of moisture-impregnated insulation. When the safe is exposed to heat the moisture is driven off as steam, allowing for dissipation of the heat. Locking devices for fire resistive and burglary resistive safes are similar: key locks, combination locks, and time locks. However, the locking device does not have anything to do with determining whether or not the unit is a fire resistive container. The construction features and performance standards of the fire resistive safe are such that very little protection is provided against the safecracker.

Once exposed to a fire, a fire resistive safe does not have the degree of protection for which it was originally rated. Exposure to heat will drive off moisture from the insulation and reduce its future protection capability.

Burglary and Robbery Resistive Safes

Burglary and robbery resistive mercantile safes, commonly known as money safes or money chests, are classified by SMNA specifications, by U.L. ratings, and by design features of the door, walls and lock, and are listed by insurance underwriters according to these specifications into their mercantile safe policies. Table 8-2 gives some of the more basic U.L.

safe classifications and Table 8-3 shows the incorporation of U.L. labeling into insurance specifications.

Table 8-2[2]
Underwriters' Laboratories
Burglary Resistive Safes.

Classification	Description	Construction
TL-15	Tool Resistive	Weight: 750 pound minimum Body: Minimum 1″ steel or equal Resistance: Door and front of unit must withstand attack with common hand and electric tools for 15 minutes.
TL-30	Tool Resistive	Weight: 750 pound minimum Body: Minimum 1″ steel or equal Resistance: Door and front of unit must withstand attack with common hand and electric tools, plus abrasive cutting wheels and power saws for 30 minutes.
TRTL-30	Torch and Tool Resistive	Weight: 750 pound minimum Resistance: Door and front of unit must withstand attack with common hand and electric tools, abrasive cutting wheels and power saws, and/or oxy-fuel gas cutting or welding torches for 30 minutes.
TRTL-30X6	Torch and Tool Resistive	Weight: 750 pound minimum Resistance: *Door* and *entire body* must withstand attack with tools and torches listed above plus electric impact hammers and oxy-fuel gas cutting or welding torches for 30 minutes.
TXTL-60	Tool, Torch and Explosive Resistive	Weight: 1000 pound minimum Resistance: Door and entire safe body must withstand attack with tools and torches listed above plus 8 ounces of nitroglycerine or equal for 60 minutes.

NOTE: Various combinations of classification ratings are available according to safe construction, design features, and performance standards.

Table 8-3[3]

Commercial Lines Manual — Insurance Service's Office
Burglary and Robbery — Resistive Equipment.

Safe, Chest or Security Locker Classifications	Construction: Doors (Combination Locked)	Construction: Walls
A	Steel less than 1″ thick, or iron	Body of steel less than ½″ thick, or iron
B	Steel at least 1″ thick	Body of steel at least ½″ thick
	Night depository - steel at least 1½″ thick	Body of steel at least 1″ thick
BR	Steel at least 1½″ thick	Body of steel at least 1″ thick
	Safe or chest bearing the label: "Underwriters' Laboratories, Inc. Inspected Tool Resisting Safe TL-15 Burglary".	
	Night Depository — Receiving safe to be equal to at least Class "BR"	
	The receiving safe and chute to be encased in at least 6″ of reinforced concrete.	
C	Steel at least 2″ thick	Body of steel at least 2″ thick
	No longer manufactured	
	Night depository - Receiving safe to be equal to at least Class "BR." Depository head to bear U.L. inspected label. Receiving safe and chute to be encased in at least 6″ of reinforced concrete.	
D	Steel at least 2″ thick	Body of steel at least 2″ thick
	When contained within a safe:	
	Steel at least 1″ thick	Body of steel less than ½″ thick or iron
	No longer manufactured	
E	Steel at least 2″ thick	Body of steel at least 2″ thick
	When contained within a safe:	
	Steel at least 2″ thick	Body of steel less than ½″ thick, or iron
	No longer manufactured	
F	At least two: Steel aggregating 5″ or more in thickness - no door less than 1″ thick	Body of steel at least 2″ thick
	No longer manufactured	
G	Round lug-type, steel at least 1½″ thick	Body of steel at least 1″ thick
	If this safe is outside of a vault, this safe is to be encased in at least 6″ of reinforced concrete and the door is to be equipped with at least a two movement timelock.	

	Safe or chest bearing one of the following labels: 1. "Underwriters' Laboratories, Inc. Inspected Tool Resisting Safe TL-30 Burglary" 2. "Underwriters' Laboratories, Inc. Inspected Torch Resisting Safe TR-30 Burglary" 3. "Underwriters' Laboratories, Inc. Inspected Explosive Resisting Safe with Relocking Device X-60 Burglary"
H	Safe or chest bearing one of the following labels: 1. "Underwriters' Laboratories, Inc. Inspected Torch and Explosive Resisting Safe TX-60 Burglary" 2. "Underwriters' Laboratories, Inc. Inspected Torch Resisting Safe TR-60 Burglary" 3. "Underwriters' Laboratiories, Inc. Inspected Torch and Tool Resisting Safe TRTL-30 Burglary"
I	Safe or chest bearing the following label: "Underwriters' Laboratories, Inc. Inspected Torch and Tool Resisting Safe TRTL-15x6 Burglary"
J	Safe or chest bearing the following label: "Underwriters' Laboratories, Inc. Inspected Torch and Tool Resisting Safe TRTL-30x6 Burglary"
K	Safe or chest bearing one of the following labels: 1. "Underwriters' Laboratories, Inc. Inspected Torch and Tool Resisting Safe TRTL-60 Burglary" 2. "Underwriters' Laboratories, Inc. Inspected Tool Resisting Safe TXTL-60 Burglary"

The following are construction features common to burglar resistive safes.

1) Safes and chests are constructed of laminated or solid steel.
2) Laminated steel is defined as two or more sheets of steel with the facing surfaces bonded together with no other material between the sheets.
3) Each door of each safe, chest, or security locker must be equipped with at least one combination lock, except a safe, chest or security locker equipped with an Underwriters' Laboratories, Inc. labeled key lock.
4) Doors. Thickness of steel is exclusive of bolt work and locking devices. If a safe has more than one door, one in front of the other, the combined thickness of the steel in the doors, excluding any door with less than 1 inch of steel, must be used in applying the following classifications.
5) Combination Lock. Each safe, chest, cabinet or vault must be equipped with at least one combination lock, except a safe or chest equipped with a key lock and bearing the label,. "Underwriters' Laboratories, Inc. Inspected KL Burglary."

The proper selection and use of a money safe will provide for the effective protection of valuables, reduce insurance premiums, increase employee efficiency and morale, and deter/prevent losses. Because money safes have unique design and construction features, the purpose and manner in which one is to be used must be given every consideration before the decision to purchase is made. For example, is the safe to be used in a jewelry store, a department store, a movie theater, etc.? What types of valuables will be stored in the safe, i.e., cash, precious gems, etc., and what will be the volume of items stored? What design features are required to incorporate the safe into reasonable and secure precedural usage? And, perhaps most importantly, what are the risk factors to the safe and its contents, e.g., employees, burglars, robbers, etc., and what is the probability of their occurrence? The decision to purchase a TL-15, TRTL-30X6, or some other classification should depend on a thorough examination of such factors.

Safecracking Methods

Most safes in use today cannot withstand the efforts of the modern-day safecracker equipped with high quality cutting and burning tools. Once the safecracker has defeated or bypassed the perimeter barriers and alarm system (if present), there are several techniques available to him to attack the safe.

Drilling or Punching

Many safe door locking mechanisms can be defeated by knocking off the combination dial or by drilling a small hole near the combination dial to expose the locking device. A safe without relocking devices, which jam the locking mechanism in place, offers little resistance to this technique.

Burning

To burn a safe is to cut an opening in the wall or door of the safe with high-temperature oxyacetylene torches or "burning bars." This attack is intended to create an opening large enough to expose the locking mechanism or to remove the contents of the safe. A burning bar or thermic lance is a device consisting of a hollow metal bar into which are packed ferrous alloy rods. To one end is attached an oxygen tank which feeds through the ferrous alloy rods. When lit at the burning end with an acetylene torch, tremendous heat is generated, capable of burning through the toughest steel safe.

Peeling

Sometimes safe doors and walls are constructed of thin steel plates laminated or riveted together to form the total thickness of the door or wall. The safecracker attacks the seams of these metal plates with pry bars and other tools to peel back the layers of metal exposing the locking mechanism or the interior of the safe.

Ripping

Ripping is similar to peeling with the difference being that ripping can be accomplished against a solid, metal walled container having a very thin outer and/or inner wall.

X-Ray

X-ray examination of the locking mechanism can reveal the position of the combination, and the manipulation necessary to open the safe. While this is not a common technique of safecracking, many safes are constructed with shields which protect the combination from the X-ray.

Explosives

The use of explosives as a means of safecracking has decreased greatly in recent years, since more efficient and safer tools and techniques are now available. However, nitroglycerin, plastic explosives and other materials can still be used to "crack" a safe.

Power Tools

Power driven rotary devices, hydraulic tools, and power drills can be utilized to pry, cut, spread, and drill openings into the door or body of the safe.

Manipulation

Other than the techniques of X-ray or manipulation to open a safe, safecracking involves forceable entry. Few safe crackers have the skill to manipulate a safe combination without some prior knowledge or condition being present. Most often, manipulation is the result of a burglar having discovered or stolen the combination to the safe.

No safe is impenetrable. However, the old axiom "you get what you pay for" is certainly applicable to safes. The burglary resistant properties provided by a safe are in direct proportion to its cost.

A safe, then, is not totally "safe," and other measures must be taken to provide for its protection. Perimeter barriers, adequate locking hardware, electronic alarm systems, and security procedures must be apparent if the desired level of security is to be accomplished.

How the safe is used is just as important as its quality of construction and performance. To hide the safe in the back room, to be careless with the control of its combination or keys, or to overdepend on its protective capability all invite theft of its contents.

Additionally, the security of a safe can be and often should be improved by taking steps to reduce its vulnerability. For example, any safe that weighs less than 750 pounds should be anchored to the building structure, thereby making its removal from the premises very difficult. A situation where this is quite common occurs when a fire resistant safe is purchased and is encased in concrete. This not only adds to its fire resistant qualities but also makes it less vulnerable to most of the safecracking techniques.

Once the decision is made to purchase either a fire resistive or burglary resistive storage container, consideration must be given to its utilization and function as a protective device. Business practices, risk factors and environmental conditions can change to the point that a safe currently in use is no longer adequate for the needed level of security.

Vaults

A vault is a room designed for the secure storage of valuables and which is of a size and shape sufficient for entrance and movement within by one or more persons. A vault is different from a safe in that it is larger, a part of the building structure, and is constructed of different materials.

Vaults should have walls, floor and ceiling of reinforced concrete at least 12 inches in thickness. The vault door should be made of steel or other torch and drill resistant material, be equipped with a combination lock, a fire lock and lockable daygate (figure 8-1). The door should also be designed to afford an appropriate degree of fire resistance (see Table 8-1).

Ratings for vaults are established by the Insurance Services Office. The ratings are based on the type of construction materials utilized, their relative thickness, construction standards, and equipment features. Materials, whether concrete, stone or steel, must be of an appropriate thickness for each rating category. Materials must be fabricated and installed according to set methods and guidelines. Construction standards must be such that the rating of the vault will not be invalidated by such variances as excessive openings in number or size. Equipment features must be met in certain rating categories, such as inclusion of combination locks and time locks (figure 8-2).

A recent innovation in vault construction and installation techniques is the modular, all metal unit. Used primarily for satellite facilities such as small branch banks, the modular vault arrives at the construction site in sections, to be lifted and fitted in place. The technique saves time and installation costs and such units are classified by Underwriters' Laboratories according to their design and construction features.

Figure 8-1
Standard Vault Door, Note Daygate Behind Vault Door.

Vault walls then can be steel, concrete, or stone. Table 8-4 is for comparison of the different kinds of wall materials.

Electronic alarm protection of vault walls, ceilings, and floors is generally accomplished by audio microphones or vibration detectors that sense attempts to penetrate the vault structure. Vault doors are protected by means of door contact switches and heat sensors. Some vault doors have a built-in wire grid that causes an alarm if the wires are cut or melted. Additional protection of the vault can be provided by area or space devices such as the motion detector, which would react to anyone moving into the area around the vault.

Information Security

Certain records and information in many organizations are absolutely

Figure 8-2
Safety Deposit Boxes Within a Vault.

Table 8-4
Vault Wall Material Equivalents.[4]

Steel Lining		Nonreinforced Concrete or Stone		Reinforced Concrete or Stone
¼ inch	=	12 inches		
½ inch	=	18 inches	=	12 inches
¾ inch	=	27 inches		
1 inch	=	36 inches	=	18 inches
1¼ inch	=	45 inches		
1½ inch	=	54 inches	=	27 inches

vital to effective continued operation. The physical destruction of such information or the loss of its confidentiality could result in one or more of the following:

1) Decreased or terminated production capability.
2) Loss of competitive edge in marketplace.
3) Inability to provide services or products.
4) Inability to satisfy certain legal requirements.
5) Complication of dealings with suppliers and customers.
6) Financial detriment to employees and stockholders.

Some of the more common records or information kept by organizations include:

Accounts Receivable	Trade Secrets
Accounts Payable	Engineering Data
Financial Reports	Process Formulas
Bank Records	Sales Records
Contracts	Tax Records
Inventory Records	General Ledgers
Legal Documents	Negotiable Instruments
Production Records	Stockholder Records
Payroll and Personnel Records	Research Data

There is really no universal method applicable to the security and protection to records of every kind of organization. Information comes in a variety of forms and its relative importance varies from organization to organization. Thus, the first stage in providing security and protection of information is to develop a procedure for evaluation and control. Generally the following four steps must be undertaken:

1) A complete inventory of records and information.
2) An appraisal of the organizational value of each type of information.
3) Development of an information classification system.
4) Application of appropriate levels of security as determined by the information classification system.

Once all the sources and types of information have been determined, a classification system must be established to differentiate between those records which are vital and those that are merely helpful. There are a number of different terms that can be used to separate and identify the various categories of information according to organizational value. One of the most common sets of terms used to establish categories of business records and information is the following:[5]

1) Vital Records. Records that are basically irreplaceable and are of the greatest value to continued operation of the organization.
2) Important Records. Records which can only be replaced at great expense of time and money.
3) Useful Records. Records which when lost would create inconveniences but could be replaced rather quickly.
4) Nonessential Records. Those records which are unessential to effective organizational operation.

Other terms that could be used to categorize records are top secret, secret, confidential, etc., but the end product would be the same, i.e., a system that implies the importance and composition of organizational information.

Methods of Protection

The protection of information can be accomplished by one or more of three methods: procedural controls, duplication, and storage.

Procedural controls are axiomatic to all areas of security and especially so in information security. The process of information classification is the initial step of establishing control of the flow of information through the organization. Procedural controls must be established to ensure that availability, responsibility, and accountability of information to personnel is restricted and based only on "the need to know." Accordingly, each organizational record should be secured and protected in descending order of importance, that is, the more vital the record and the information contained, the greater the degree of control and security.

Duplication of records is both a security risk and a security asset. It provides for needed dissemination of information throughout the organization, but it also provides a means by which information can be copied surreptitiously and made available to others outside the organization. Thus, procedural controls must be adopted to ensure that unwarranted duplication does not occur. However, duplication of records can be used to provide an essential element of providing additional security. Any record which is vital or important to the organization can be duplicated and dispersed to one or more remote locations away from the primary storage. In the event that the primary source of this information is lost, the organization can continue without interruption.

The manner in which information is stored is a determining factor in the effective level of security. Information stored in desks and ordinary file cabinets has minimum protection against forcible entry, whereas information kept in a safe or vault is much more secure. That information which is declared to be important or vital should be protected by an appropriate security storage container. A careful analysis of risk factors must be done in order to decide the type of protection needed. The kinds of threats that must be considered include fire, espionage, and theft.

Computer Security

Computer storage of information has become commonplace in business, industry and government. Computers have greatly increased the ability to store, retrieve, manipulate, and transmit vital information. Yet the misuse, damage, or loss of a computer can render helpless or destroy an entire company. Given this destructive potential, electronic data processing systems must be afforded an appropriate high level of security. Access to and operation of the computer unit must be controlled, and any attempt to enter, manipulate or otherwise obtain information must be preventable and detectable.

Storage of data and programs on magnetic tapes, cards, discs, or drums is a vital part of any computer operation. The capacity to store such a large amount of data on such a small physical device is both a liability and an asset. Steps must be taken to provide security for the physical storage of tapes, cards, discs, etc. Special storage units for the various data forms are available which provide not only security but also a controlled environment. Temperature and humidity of the storage unit or tape library must be within acceptable limits. The range of risks involving computers include fire, espionage, sabotage, accidental losses, theft, fraud, embezzlement and natural disasters.

Summary

The security requirements of each facility must be individually analyzed in light of physical factors, type and extent of risk exposure, and business experience. Whether the center of concern is a safe or a vault, it is security in depth that decreases the chances for criminal success.

Information is a primary asset and must be protected as such. Only after an in-depth evaluation of all records can an effective and appropriate level of security be accomplished for the control of the flow and storage of information.

Discussion Questions

1. What are the two types of safes and their purposes?
2. Describe the circumstances which may dictate the use of a maximum fire protection safe.
3. Discuss the differences between safes and vaults.
4. Outline the steps taken to protect information.
5. Describe the three methods of protecting information.

Notes

1. Robert C. Holcomb, "Rating of Safes: Part 1," *Keynotes* (December, 1977): 17.
2. Adapted from Lawrence J. Fennelly, *Handbook of Loss Prevention and Crime Prevention.* Butterworth Publishers, 1982.
3. Adapted from *Commercial Lines Manual.* Insurance Services Office, 1985.
4. *Ibid.*
5. Raymond M. Mombossiee, *Industrial Security for Strikes, Riots and Disasters.* Charles C. Thomas, 1968.

Chapter 9
Security Personnel

Private security personnel now outnumber public law enforcement officers in the United States. There are a number of reasons for this: the general economic growth of the postwar period; the movement of population from the central cities to the suburbs, which has led to the development of shopping centers and malls; tax restraints, such as Proposition 13 in California and the Gramm-Rudman-Hollings Act, which often reduce the funds available for public law enforcement; and an increasing fear of crime which has caused Americans to demand more security.

Thus, because of the importance of private security, there is a need for a thorough understanding of the processes that must be utilized to maximize the efficiency and effectiveness of private security personnel. At this point, there is no need to differentiate between contract and proprietary security services, because both must deal with the same problems of recruitment, training, etc. The issue of contract vs. proprietary personnel is a real and controversial one, however. For example, in some states, legislation provides one set of selection and training standards for contract security and another for proprietary. Strangely enough, this is often accomplished through "non-legislation" as it relates to proprietary security personnel. The legislatures simply do not make the legislation binding on proprietary services. Contract security companies have argued, with little success to date, that this discriminates against them by reducing their opportunity to compete in the market place in trying to provide their services to a company that presently has a proprietary force. Obviously, although most security personnel will publicly deny it, the real issue seems to be "turf protection" for proprietary security. As long as the contract companies have legislation that costs them money, they will be less capable of taking over security services at companies that presently have proprietary forces. To the new person in the field of private security this issue of separate legislation may seem minor, but to those who have experience in private security, the issue is real and important.

This chapter is divided into six sections: current industry criticism, recruitment, selection, training, licensing and registration, and the duties of private security personnel.

Current Industry Criticism

Most private security personnel will readily admit that the industry is criticized and that much of that criticism is warranted. However, criticism is also leveled at public law enforcement, courts, corrections, and almost all other organizations and services which deal with the public. Certainly the professions of teaching, law, and medicine have been increasingly criticized recently. Thus, private security personnel should not be overly defensive in their response to criticism.

Most recent studies conducted of the age, race, education, general physical characteristics, etc. of private security personnel have found that the differences between private and public law enforcement, *on these factors*, are minimal. The only exception to this was a study conducted in the early 1970's by the Rand Corporation. However, it should be noted that the Rand Study was done with a small sample and in one geographic location (southern California), and it is generally agreed that their sample was not typical of the industry as a whole. It is interesting to note that the Rand Study is more often cited in the literature than the other research that has been conducted.

The problem lies, if we accept the statement that differences are minimal as outlined above, in the factors that cannot be quantified. These factors include motivation, dedication, common sense, and many psychological factors. For example, it is fairly well accepted that persons sometimes enter the field of private security because of what is called the "John Wayne Syndrome." The opportunity to carry a gun and be recognized as a powerful person has motivated some people to enter the field of private security. It is also known that the same motivation has caused persons to apply for public law enforcement positions. The difference is that, for the most part, public law enforcement uses testing and referrals to psychologists and psychiatrists to try to determine the applicants' thoughts and perceptions about these matters. Private security firms, on the other hand, almost never take the time, effort, and money to review applicants' feelings about this issue.

Anthony Potter, at the First Annual Conference on Private Security conducted at the University of Maryland in December, 1975, provided good insight into the subject by the concept of the "Private Security Vicious Circle," which is depicted in figure 9-1. Most persons involved in the private security industry see few positive steps being taken to help people "break out" of the circle.

Recruitment

Newspaper advertisements and listings of contract security companies in the Yellow Pages, at present, provide the potential employee with the best

Figure 9-1
Private Security Vicious Circle.

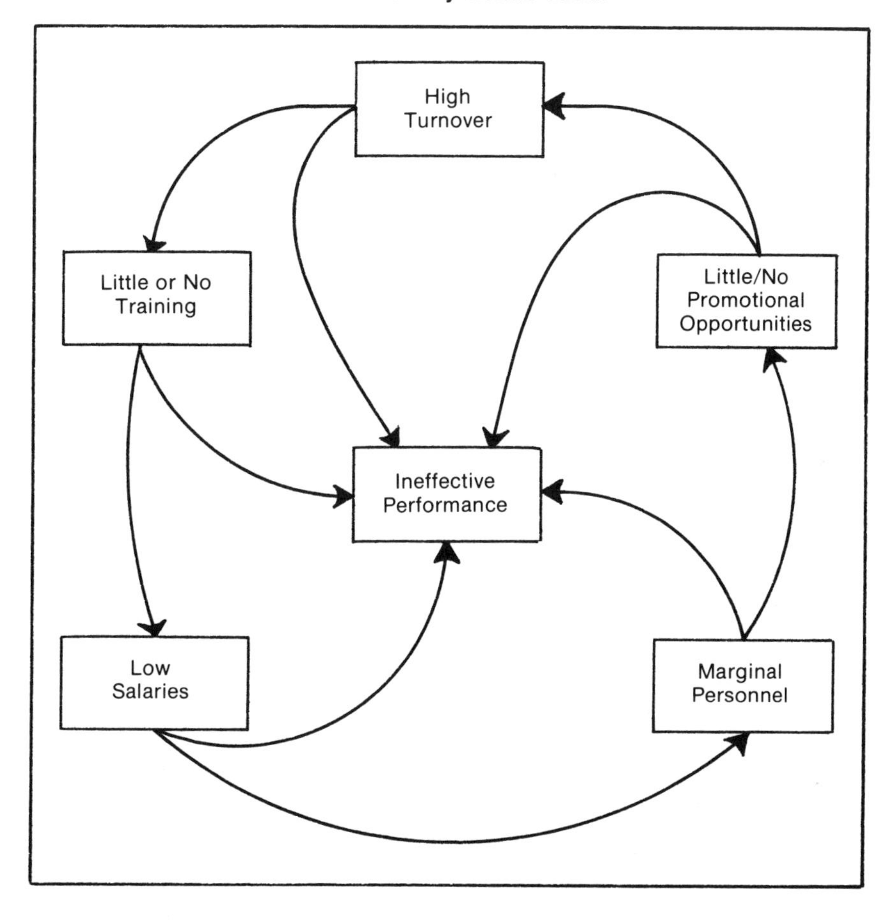

leads to employment opportunities. Almost every day, in papers throughout the United States, there are ads placed by companies seeking applicants. Because of the low salaries in most private security jobs, these ads are seldom large. Furthermore, careful reading of the ads will clearly show that the qualifications are, at best, minimal. Traditional qualifications for employment, such as education and experience, are seldom listed. About the only general qualification is a "clean police record," which certainly should be expected. Unfortunately, there is often little agreement among employees about what is considered a "clean record," and what is acceptable to one employer might not be acceptable to another. Most security firms in major cities maintain full-time employment offices. However, the opportunities are somewhat limited for potential employees from other areas. The larger contract companies and larger corporations do, from time to time, send recruiters to the non-metropolitan areas, but this is usually as a last resort when they cannot get applicants locally. Thus, private security employees, for the most part, come from areas close to where they live. From a recruitment standpoint this is not much of a disadvantage, since most employers are happy to get any type of person and are not overly concerned with their qualifications.

The recruitment of management personnel follows a somewhat different process. Companies fill these positions by promoting present employees or by recruiting persons from the outside. Two major sources of recruitment are heavily utilized. First, the companies actively maintain contacts with public law enforcement agencies through personal contacts and through membership in and attendance at meetings of local public law enforcement associations. For example, it is not uncommon for a representative of private security to make an announcement of a position opening at a meeting, and to promote the position informally before and after the meeting.

The second source used is campus recruitment. In some cases the recruitment is done by the security department, usually in coordination with the criminal justice faculty. In many cases, however, the companies simply utilize the college recruitment processes of the company and do not make any special effort to recruit security personnel independently.

In summary, the recruitment of security personnel is not a complex or sophisticated process. The high turnover rate (over 100% per year in many cases) is one of the main problems, and this is clearly reflected in the constant advertising done by the employers. Management positions are filled by a somewhat different procedure, and it will take more initiative on the part of applicants since the recruitment of retiring police officers and college recruiting are the main methods now being used. Individuals who do not have the experience or opportunity to learn of openings through these processes will have to send resumes, visit employment offices, etc. if they want to obtain management positions.

Selection

There are no selection standards that apply to the entire private security industry. Extensive research has been conducted on this subject and the conclusions are always the same—no uniformity exists. Many government studies and writers on the subject have indicated that there should be minimum qualifications for selection, similar to those used in public law enforcement. (There is some disagreement as to whether the qualifications should be the same as for public law enforcement.) Unfortunately, there is little hope that this will occur, for several reasons. First, private security is a competitive business, and any set selection standards would tend to reduce the pool of, and increase competition for, available applicants. Security employers fear that this will raise salary levels and make it harder for them to sell their services in the market place. Second, unless all private security employers agreed to the selection standards (which is highly unlikely), the only recourse would be legislation, controlled through licensing and registration by the state. Most companies view this as "interference" and would oppose it. No other pressure group is likely to push for this legislation, so it seems that no comprehensive selection standards will evolve through the legislative process.

One government study dealt with the issue of "minimum pre-employment screening qualifications" in some detail, and the pertinent portions are worthy of review. This material was developed by the Private Security Task Force to the National Advisory Committee on Criminal Justice Standards and Goals during their work in 1975 and 1976. The Task Force was made up primarily of security professionals and some public criminal justice professionals. The National Advisory Committee was made up from a more general selection of criminal justice professionals and elected officials. Even these two groups could not agree on minimum educational qualifications (see discussion of educational requirements in Appendix A, Standard 1.8).

The screening process is another point in the selection of personnel. One of the most crucial problems in the recruitment of private security personnel is the interview-today-start-work-tomorrow syndrome. Unfortunately, many applicants begin their employment 24-48 hours after application. Companies defend this process by insisting that they will observe the employee's performance carefully and that the best judgment of their qualifications will be their work performance. For those familiar with the private security industry, this rationale is weak. The truth is that most companies are unwilling, partially because of the turnover rate, to invest the time and money to thoroughly screen applicants. The following recommended procedures are considered minimal, but are still the exception rather than the rule in the private security industry. The following selected portions of Standard 1.3 of the *Report* provide a good overview.

Screening Interview

The screening interview is a two-way communication—employer to applicant and applicant to employer. Although somewhat subjective, it allows both parties to assess the job situation. Employer questions should include:

1. Why do you want the job?
2. What are your career objectives?
3. What interests you about the job?
4. Other job-related questions.

The employer should clearly indicate to the applicant the requirements, positive and negative aspects, salary and fringe benefits, and other pertinent factors about the job.

The screening interview also allows the employer to assess the applicant's character. Although such an assessment is admittedly highly subjective, the applicant's demeanor and attitude during the interview may indicate the need for more careful background investigation or even psychological testing.

Honesty Test

For the purpose of this standard, honesty tests refer only to written tests that allow employers to gain insights into a prospective employee's honesty without extensive costs. In general terms, honesty tests are designed to measure trustworthiness, attitude toward honesty, and the need to steal.

Several paper-and-pencil honesty tests were reviewed for this report. This independent evaluation determined that the tests appear to have high face validity. Several validity and reliability studies supporting such tests have been published in scholarly journals. However, it should be noted that much of the supporting evidence is based on subsequent detection-of-deception examinations of persons who had taken the written honesty tests. Nevertheless, honesty tests used with background investigations should furnish a reliable method of determining honesty.

Background Investigation

Background investigations should be conducted prior to employment and/or assignment. . . .

Too often, employers do not conduct any background investigations or investigations are sketchy. Many employers use only the telephone and/or form letters for background information. Such methods do not provide sufficient data for effective verification and evaluation. Although costly, field investigations should be encouraged to provide valuable information about an individual's character and ability that cannot be gained by other means.

Other Screening Considerations

Job-related psychological tests and detection-of-deception examinations are additional processes that could be included in the screening process. Job factors such as access to funds and other property, control of personnel, whether armed, and so forth, should determine the types of job-related tests that can best serve the employer and the public. It is shocking to realize that many armed guards are not screened to determine if they have major psychological problems that would clearly render them unacceptable for employment involving carrying a deadly weapon. Obviously, extreme care should be taken to ensure that all screening measures are job-related and are not an invasion of the applicant's individual rights. It is also important that all screening methods be administered and evaluated by competent personnel and the results carefully protected from illegal release.

Properly conducted, pre-employment screening will aid employers in selecting capable and trustworthy employees. By eliminating those unsuited for private security work, such screening processes also will lead to increased productivity and lower turnover rates.

The establishment of pre-employment screening qualifications and the screening process are probably the most critical issues in the process; however, there are additional issues. One of these items is the exchange of information between employers. At first glance this would seem to be quite simple, but it is not at all. For example, Company A might have a "bad" employee who is either fired or resigns. Shortly thereafter Company B calls Company A to verify employment. Company A is aware that Company B has a contract with the XYZ Company and would like to have that contract for themselves. There is now a strong (though wrong) incentive for Company A to give a good recommendation for the employee, in hopes that he will be hired, perform poorly and thus get the XYZ Company dissatisfied with Company B, and get the bid themselves. However, it is in the best interests of private security generally for the employers to exchange information on at least the following items:

Dependability
Honesty
Initiative
Judgment
Loyalty

In addition, as permitted by law, the following additional information should be exchanged.

1) Arrest and/or conviction information,
2) Use of drugs and/or alcohol,
3) Poor interpersonal relationships with clients or fellow workers,
4) Poor credit rating,
5) Improper use or abuse of authority and/or force, and
6) Psychological unsuitability.

Another issue is affirmative action and equal employment opportunity practices. Most government contracts require this in the selection process, but many smaller firms have no government contracts and are difficult to monitor in regard to these issues. Studies by the Private Security Task Force in New Orleans, Louisiana and St. Louis, Missouri in 1975 indicate that, in comparison to race ratios in those cities, private security has a proportionate ratio of minority employees.

The most recent comprehensive study on this issue was conducted by Hallcrest Systems, Inc. of McLean, Virginia and reported in *Private Security and Police in America*. This book is based on a 30-month descriptive research project under a grant from the National Institute of Justice, U. S. Department of Justice. One interesting aspect of this information is that it provides data on both proprietary and contract security personnel. While many people continue to believe that older persons are the main source for private security employment, this is not true. The median age for both groups in this study was 31-35 years of age. Also, 95% of the proprietary and 90% of the contractual personnel had at least a high school degree or GED.

A factor that has become very apparent, especially in the last decade, is that private security is making greater use of females than public law enforcement. Separate studies conducted by Professor Jack Molden and Assistant Professor Michael Hoefling of the University of Illinois Police Training Institute reveal that 5% of police personnel in Illinois are female. A study reported in the Hallcrest Report indicated 6% of public law enforcement officers were female. Other studies in the Hallcrest Report revealed that 24% of the employees in proprietary companies and 12% in contractual companies were female. This is especially significant because there are approximately 450,000 more employees in private security than in public law enforcement.

Another problem is that various states have developed rules and regulations regarding who has access to their Criminal Justice Information Systems and, for the most part, private security employers are denied access. This puts a serious restraint on the ability of private security employers to assess the applicant's qualifications. Many employers use an informal system, through cooperation with various public agencies, to "get around" this restriction. However, this process, in addition to being illegal, does not meet the needs of the employers. One security executive has indicated his total frustration with the situation by this example. His company does some government contracting which requires security clearances for certain employees, and his company is responsible for conducting the screening and selection of these persons. He does not have official access to the necessary information, and so has to, in effect, violate one law in order to comply with another.

Training

There is probably no more important issue in private security today than the training (or rather the lack of it) received by private security personnel. All who have studied this issue have concluded that there is a lack of realistic and viable training for operational level personnel. There have been, however, improvements in the training of supervisory and management personnel in recent years, and this issue will be discussed more thoroughly in Chapter 15. In this chapter the discussions will center on the training of operational personnel.

There are a number of major differences between private and public law enforcement operations, but the differences are most glaring on the issue of training. Today, almost all of the states have adopted comprehensive training programs, especially for new officers, who are employed in the public sector. Most of these programs require a minimum of 400 hours of entry-level training and many states exceed that amount. Improvements have been occuring in relation to private security training. Studies conducted by the Private Security Task Force in 1975 revealed that seven states had training requirements. An updated study reported in the Hallcrest Report indicated that sixteen states now have training requirements. Even more important is the perception of training by security employees. Seventy percent of the proprietary and 65% of the contractual employees expressed the opinion that their training was adequate when responding to a survey sent as part of the research conducted by Hallcrest Systems, Inc. of McLean, Virginia.

The key distinction that must be made in discussing private security training is between armed and unarmed personnel. At least at the philosophical level there is agreement, even among private security executives, regarding the need for training of armed personnel. Although the firearm may never have to be used, its misuse can cause difficult legal consequences, injury, or death.

The material in Appendix A from the Task Force *Report* was prepared after, and as a result of, extensive dialogue between the Task Force and the private security industry. Probably no other portion of the *Report* was debated more intensely and with more emotion. Originally, the staff of the Task Force recommended a 40-hour pre-assignment training course, but through the public hearing process this was reduced to 8 hours formal preassignment training and 32 additional hours of training within 3 months of assignment. It should also be noted that 16 of the 32 hours can be supervised on-the-job training. Since Standard 2.5 is the most thorough and detailed curriculum and discussion of the issue that is available, it is reproduced in Appendix A.

As noted earlier, there is some agreement on the issue of training for armed personnel. Though there was disagreement as to the amount of training while the Private Security Task Force was conducting public hearings, the end result of that process was accepted by the Task Force and the private security industry representatives with little debate or change. It closely follows the 24-hour training program designed by the Illinois Local Governmental Law Enforcement Officers Training Board in 1976 for the training of full-time and part-time public law enforcement officers in Illinois. It should be carefully noted that it involves both training prior to a job that requires a firearm *and* requalification at least once every 12 months. It should be noted that the requalification requirement *exceeds* that of many public law enforcement regulations or statutes. Even the Illinois requirements do not have a requalification provision. However, the State of Illinois, in 1985, increased the firearms training program to 40 hours, but it still does not have a requalification program.

The amount of time devoted to firearms training of private security personnel continues to be a problem. A survey of security employees in Baltimore County, Maryland and Multnomah County, Oregon metropolitan areas revealed that the median hours of firearms training for proprietary personnel was 12 hours and 8 hours for contractual employees. The averages were 16.7 hours and 12.7 hours for proprietary and contractual, respectively.

From these figures it is apparent that adequate time for firearms training in private security has not been attained. While there appears to be some improvements in the last decade, such as the 30-hour training requirement in Texas, there is still a long way to go before most of the firearms training in private security will be realistic in terms of the safety of the trainees and the citizens they may contact while performing their security functions.

The research reported in the Hallcrest Report and presented in Table 9-1 would tend to indicate that the model training courses developed by the Private Security Task Force have been used as a guide for private security training. Increased training is a positive step toward attacking the private security vicious circle described earlier in this chapter.

Licensing and Registration

Thus far in this chapter we have discussed the recruitment, selection, and training of private security personnel. These activities are important steps toward improving private security services. However, there is another step in the process: registration of the individual employees with a state regulatory agency or board.

Even though businesses generally oppose any type of additional government control, those in private security recognize the need for some type of government regulation to insure uniformity of requirements. Though the concept of regulation is accepted, there is no unanimity on the issue of who should be the "regulator". There is, however, general agreement that it should not be the same agency or board that regulates public law enforcement. There is general agreement that private security represen-

Table 9-1
Training Subjects Reported By Security Employees.

TRAINING SUBJECT	PROPRIETARY	CONTRACT GUARDS	ALARM RUNNERS
Fire Protection & Prevention	76%	81%	82%
First Aid	69%	48%	55%
Legal Powers: Arrest, Search, Seizure	83%	74%	55%
Investigation & Detention Procedures	75%	62%	64%
Firearms (classroom)	13%	10%	64%
Firearms (firing range)	21%	7%	64%
Building Safety	61%	69%	64%
Crisis Handling	51%	55%	27%
Crowd Control	45%	41%	18%
Equipment Use	64%	48%	82%
Report Writing	82%	79%	73%

Source: Site Surveys of Security Employees, Baltimore County, Maryland and Multnomah County (Portland), Oregon metropolitan areas, Hallcrest Systems, Inc., 1982.

tatives should, along with *some* public law enforcement officials, other government officials and citizens, be involved in the establishment and enforcement of the registration issues. For readers interested in a detailed study of this matter, the *Report of the Task Force on Private Security* (1977) provides an in-depth review of the issue and detailed recommendations.

For a better understanding of the issue of licensing and registration in this chapter several terms must be defined.

Licensing— The act of requiring permission from a competent authority to carry on the business of providing security services *on a contractual basis.*

Registration—The act of requiring permission from a state authority before being employed *in the private security industry.*

The italicized portions of the above definitions highlight the differences. Licensing does not, and in the opinion of many should not, apply to proprietary security services since the companies that operate them are already licensed, in most cases, by some government authority to produce a product or sell a service. Specific licensing of proprietary security personnel would, in their opinion, be unnecessary. On the other hand, the individual employees in both proprietary and contract services should be registered to ensure uniformity.

As noted earlier in this chapter, there have been significant changes in the number of females and increased training for private security personnel during the last ten years. This is not true with regard to licensing and registration. There have been relatively few attempts to establish licensing and registration in additional states although there have been modifications made in some states. Thirty-five states license guard and patrol companies, but only 22 states and the District of Columbia require the registration of guards.

Once again a distinction can and should be made between armed and unarmed personnel. Further, it is generally conceded that there should be both temporary and permanent permits. The temporary permit should be for a limited duration, for example 30 days, and require at least a check of the applicant with local law enforcement agencies and other available sources. The primary responsibility for this action rests with the employing agency. The permanent registration permit should be issued by the state agency, and the Private Security Task Force established two Standards (11.2 and 11.3) which provide guidance on this issue. (See Appendix A.) It should be noted that these standards generally agree with the recommendations in this chapter regarding selection and training of private security personnel.

National Overview

There is no uniform or cohesive system of governmental licensing and registration. Charles Buikema and Frank Horvath, in an article entitled "Security Regulation — A State-by-State Update" which was published in the January 1984 issue of *Security Management*, provide information that reflects the lack of uniform or cohesive government licensing and registration. They reported their research in three (3) services (duties performed): security guards services, both contract and proprietary; alarm system contractors, and private detective services. Some of the summaries that can be made from their study are:

- 37 of the 47 states that responded to the survey (Oklahoma, South Dakota, and Tennessee did not respond) regulate at least one of the services.
- 33 states regulate contract security and guard services, but only 3 regulate proprietary guard services.
- 33 states regulate private investigators.
- 8 states regulate alarm system contractors.
- 19 states regulate minimum age of 18 for employment, 7 a minimum age of 21 for employment, 1 a minimum age of 22 for employment, and 7 a minimum age of 25 for employment.
- 16 states require a minimal level of training for armed personnel.
- 4 states require training of unarmed personnel.
- 10 states require a state fingerprint check, but only 5 require a national FBI check.

> These summaries reflect the lack of uniformity and cohesiveness in attempts to regulate the private security industry. The situation is compounded for security operations that are located in a bi-state area such as Chicago, Illinois/Gary, Indiana, or St. Louis, Missouri/East St. Louis, Illinois.
>
> There has been very little discussion and no serious proposals for any form of national licensing or registration for the private security industry. For the last two decades the major focus has been on state licensing and registration, but municipal and county licensing and registration exists in some parts of the country.
>
> In the above summary statement it can be said that since 1984 no significant changes have occurred at the state level relative to licensing and registration of security personnel. The pace of change is slow and perhaps will increase in tempo as the threat of civil liability becomes even more menacing.

Duties Performed

So far in this chapter we have addressed the issue of private security personnel from initial contact to hiring. At this point we will discuss what the job duties are after the employee has been hired.

This is a rather complex problem, because of the wide variety of duties and functions performed by employees in the broad category of "private security personnel." A reasonable starting point is the approach taken by the Private Security Task Force in emphasizing the need for job descriptions for private security personnel. This process, which unfortunately is not common in private security, is an important step in clarifying the duties to be performed. Standard 2.3 of the Private Security Task Force Report provides a concise analysis of the issue.

> As a general guide, the data recorded in job descriptions should relate to two essential features of each position: (1) the nature of the work involved, and (2) the employee type who appears best fitted for the position.
>
> With respect to the nature of the job, the following data should be included:
>
> 1. The job title;
> 2. Classification title and number, if any;
> 3. Number of employees holding the job;
> 4. A job summary, outlining major functions in one to three paragraphs;
> 5. A job breakdown, listing the sequence of operations that constitutes the job and noting the difficulty levels;

Table 9-2¹
Selected Factors
States Requiring Licensure of Guard,
Investigator, and Alarm Systems and Major Qualifications

	AK[a]	AZ	AR	CA	CT	DE	FL	GA	HI	IL	IN	IA
Areas Regulated												
Contract guards	X	X	X	X	X	X	X	X	X	X	X	X
Proprietary guards								X				
Private investigators		X	X	X	X	X	X	X	X	X	X	X
Alarm system contractors			X	X								
Licensure Requirements												
Minimum age		C	D	C	F	F	C	C	E	D	D	C
Training	X	X	X	X		X	X	X		X		
Miscellaneous												
Fingerprint check	X						X	X				

	KS	KY	ME	MD	MA	MI	MN	MT	NE	NV	NH	NJ
Areas Required												
Contract guards			X	X	X	X	X	X		X	X	X
Proprietary guards		X										
Private investigators	X		X	X	X	X	X	X	X	X		X
Alarm system contractors						X		X				
Licensure Requirements												
Minimum age	D	D	C	F	F	F	C	C	D	D		F
Training	X											
Miscellaneous												
Fingerprint check							X					

	NM	NY	NC	ND	OH	PA	RI	SC	TX	UT	VT	VA
Areas Required												
Contract guards	X	X	X	X	X	X		X	X	X	X	X
Proprietary guards								X				
private investigators	X	X	X	X	X	X		X	X		X	X
Alarm system contractors			X				X		X			
Licensure Requirements												
Minimum age	C		C	C		F	C	C	C	C	C	C
Training			X			X		X		X	X	X
Miscellaneous												
Fingerprint check		X				X	X				X	X

	WV	WI
Areas Regulated		
Contract guards	X	X
Proprietary guards		
Private investigators	X	X
Alarm system contractors		X
Licensure Requirements		
Minimum age	C	C
Training		X
Miscellaneous		
Fingerprint check	X	

[a]—Standard used by the US Postal Department for abbreviations of states.
B—Time limit exemption or other qualifying expression.
C—18 years old.
D—21 years old.
E—22 years old.
F—25 years old.
G—Education allowable substitute for experience.

6. A description of equipment used;
7. A statement of the relationship of the job to other closely related jobs;
8. A notation of the jobs from which workers are promoted and those to which workers may be promoted from this job;
9. Training required and usual methods of providing such training;
10. Amounts and types of compensation;
11. Usual working hours; and
12. Peculiar conditions of employment, including unusual circumstances of heat or cold, humidity, light, ventilation.

With respect to the employee, the data generally available should include:

1. Necessary and special physical characteristics;
2. Necessary physical dexterities;
3. Emotional characteristics, such as disposition, mood, introversion, or extroversion;
4. Special mental abilities required;
5. Experience and skill requirements.

This outline is not intended to be all inclusive but is presented to highlight the depth to which job descriptions should be prepared if they are to be effective. It was noted, however, that some job descriptions reviewed included nonsecurity functions, such as running errands. This practice should be discouraged, because it detracts from the overall effectiveness and morale of private security personnel.

In summary, the preparation of high quality job descriptions is a critical step in the personnel selection, assignment, and training processes. Without job descriptions, the employer, employee, and person responsible for developing training programs are at a tremendous disadvantage. Further, the need to relate training to the job is vital if training is to carry more significance than mere hours spent sitting in a classroom.

One of the major difficulties in discussing the issue of duties performed by private security is the complexity of the issue. For example, the General Services Administration employs its own guards and at the same time contracts for similar services. While the duties performed may be very similar, there is wide misunderstanding as to the classification of the employees. In some cases they are federal employees, and in some cases they are federal contract employees. Some people try to make a thin-line distinction and classify the latter category as "quasi-government employees."

Another example is operations such as transit authorities. In some cities, such as Chicago, the personnel in these operations are defined by the legislature as public officers, while in other cities they perform the same duties but are identified as private security personnel. Probably the best way to resolve this issue is to not be concerned with the duties they perform, but rather to classify them by the source of their funds. If they are paid directly by public funds they are "public" and if they are paid with government funds through a contract they are "private". Regardless of the source of funds we can broadly classify the duties performed into four areas.

Guard and Patrol Services and Personnel

Guard and patrol services include the provision of personnel who perform the following functions, either contractually or internally, at such places and facilities as industrial plants, financial institutions, educational institutions, office buildings, retail establishments, commercial complexes (including hotels and motels), health care facilities, recreation facilities, libraries and museums, residence and housing developments, charitable institutions, transportation vehicles and facilities (public and common carriers), and warehouses and goods distribution depots:

- Prevention and/or detection of intrusion, unauthorized entry or activity, vandalism, or trespass on private property;
- Prevention and/or detection of theft, loss, embezzlement, misappropriation or concealment of merchandise, money, bonds, stocks, notes, or other valuable documents or papers;
- Control, regulation, or direction of the flow or movements of the public, whether by vehicle or otherwise, to assure the protection of property;
- Protection of individuals from bodily harm; and
- Enforcement of rules, regulations, and policies related to crime reduction.

These functions may be provided at one location or several. Guard functions are generally provided at one central location for one client or employer. Patrol functions, however, are performed at several locations, often for several clients.

Investigative Services and Personnel

The major services provided by the investigative component of private security may be provided contractually or internally at places and facilities, such as industrial plants, financial institutions, educational institutions, retail establishments, commercial complexes, hotels and motels, and health care facilities. The services are provided for a variety of clients, including insurance companies, law firms, retailers, and individuals. Investigative personnel are primarily concerned with obtaining information with reference to any of the following matters:

- Crime or wrongs committed or threatened;
- The identity, habits, conduct, movements, whereabouts, affiliations, associations, transactions, reputation, or character of any person, group of persons, association, organization, society, other group of persons or partnership or corporation;
- Pre-employment background checks of personnel applicants;
- The conduct, honesty, efficiency, loyalty, or activities of employees, agents, contractors, and subcontractors;
- Incidents and illicit or illegal activities by persons against the employer or employer's property;
- Retail shoplifting;
- Internal theft by employees or other employee crime;
- The truth or falsity of any statement or representation;
- The whereabouts of missing persons;
- The location or recovery of lost or stolen property;
- The causes and origin of or responsibility for fires, libels or slanders, losses, accidents, damage, or injuries to property;
- The credibility of informants, witnesses, or other persons; and
- The securing of evidence to be used before investigating committees, boards of award or arbitration, or in the trial of civil or criminal cases and the preparation thereof.

Detective or investigative activity is distinguished from the guard or watchman function in that the investigator obtains information; the guard or watchman usually acts on information (or events).

Alarm Services and Personnel

Alarm services include selling, installing, servicing, and emergency response to alarm signal devices. Alarm devices are employed in one of four basic modes: local alarm, proprietary alarm, central station alarm, or police-connected alarm. Alarm signal devices include a variety of equipment, ranging from simple magnetic switches to complex ultrasonic Doppler and sound systems. Various electronic, electromechanical, and photoelectrical devices and microwave Dopplers are also utilized.

Alarm personnel include three categories of employees: alarm sales personnel, alarm systems installers and/or servicers, and alarm respondents. Those persons in alarm sales engage in customer/client contact, presale security surveys, and postsale customer relations.

Alarm installers and servicers are trained technicians who install and wire alarm systems, perform scheduled maintenance, and provide emergency servicing, as well as regular repair of alarm systems. (Alarm installers and servicers may be the same depending on the employer.) Alarm respondents respond to an alarm condition at the protected site of a client. The alarm respondent inspects the protected site to determine the nature of the alarm, protects or secures the client's facility for the client until alarm system integrity can be restored, and assists law enforcement agencies according to local arrangements. The alarm respondent may be armed and may also be a servicer.

Armored Car and Armed Courier Services and Personnel

Armored car services include the provision of protection, safekeeping, and secured transportation of currency, coins, bullion, securities, bonds, jewelry, or other items of value. This secured transportation, from one place or point to another place or point, is accomplished by specially constructed bullet-resistant armored vehicles and vaults under armed guard. Armed courier services also include the armed protection and transportation, from one place or point to another place or point, of currency, coins, bullion, securities, bonds, jewelry, or other articles of unusual value. Armed courier services are distinguished from armored car services in that the transportation is provided by means other than specially constructed bullet-resistant armored vehicles. There are also courier service companies that employ non-armed persons to transport documents, business papers, checks, and other time-sensitive items of limited intrinsic value that require expeditious delivery.

The major distinction between the services provided by armored cars and armed couriers and those furnished by guards and watchmen is liability. Armored car guards and armed couriers are engaged exclusively in the safe transportation and custody of valuables, and the firms providing these services are liable for the face, declared, or contractual value of the client's property. These service companies are bailees of the valuable property and the guards and couriers are protecting the property of their employer. This liability extends from the time the valuables are received until the time a receipt is executed by the consignee at delivery. Except for war risks, the armored car company is absolutely liable for the valuable property during such protective custody. Conversely, guards, watchmen, and their employers do not assume comparable liability for the property being protected.

As stated at the beginning of this chapter the number of persons employed in private security outnumber the persons employed in public law enforcement. Thus, the greatest improvements in the services provided by private security will be realized by proper recruitment, selection, training, licensing and registration, and performance by the personnel employed. This chapter has identified the issues and provided plans for the improvement of security personnel.

Discussion Questions

1. Discuss the "Private Security Vicious Circle".
2. List and discuss some procedures that can be used in the selection process.
3. Discuss the issues in private security training.
4. Define and discuss the issues related to licensing and/or registration of private security personnel.
5. List the four major categories of duties performed by private security personnel, and give two examples of types of duties performed for each category.

Notes

1. 1984 Copyright by the American Society for Industrial Security, 1655 North Ft. Myer Drive, Suite 1200, Arlington, VA 22209. Reprinted by permission from the January/1984 issue of *SECURITY MANAGEMENT* magazine.

Chapter 10
Risk Analysis and Security Surveys

Every business endeavor, whether it is run by an individual entrepreneur or a conglomerate, faces daily risks to its well-being and survival. Top management is ultimately responsible for organizational security. It must thoroughly investigate the policies and procedures of the organization, the conditions of the physical plant, the backgrounds and attitudes of the employees, and the relationships of the organization with the rest of the world before any worthwhile efforts can be undertaken to make the organization more secure from the threat of losses due to injury, death, damage, or destruction. In this respect, the risks discussed here are actual hazards or threats by natural or man-made forces and do not include the risks inherent with competitive business ventures that are conducted within legal bounds. It should also be understood that, although top management is responsible for security, this authority is usually delegated to another individual, i.e., the security director or manager within the organization.

Physical security can be provided for any size organization as long as it has personnel, real property, a service, or an end product to protect. An organization can be a one-chair barbershop, a municipal school, or a corporation with personnel employed nationwide and internationally. Differences in physical structure, product or service, number of employees, or local environmental factors mean that each organization must protect itself from different kinds of threats, but the basic principles of security are applicable to organizations of any size or type.

This chapter will outline the processes undertaken to conduct a risk assessment, establish loss probability, and design surveys appropriate for various types of organizations. The section concerning cost benefit analysis will explain why the employment of the most sophisticated and complex security equipment and security personnel may not be financially appropriate or strategically necessary to safeguard every facility or activity. Finally, there will be some suggestions for the development and maintenance of support for the security plans and programs by the operational and management personnel.

There are different purposes for risk assessments of different types of organizations, activities, or facilities. Basically, the risk assessments are made to determine the degree of exposure to hazards or dangers to personnel or property, as described in Chapter 3. Although some elements are similar, a risk assessment conducted to establish the security of a facility from a natural disaster would not satisfy the requirements of a physical security or crime prevention survey. The differences of these types of assessments or surveys will be explained throughout this chapter.

It is also important to understand the terminology used by personnel throughout the security industry. In the previous paragraph one can see the synonymous use of the words assessment, analysis and survey. Although they are not precisely synonymous, those words as well as audit, evaluation, inspection, investigation, measurement, and study are all used to describe the activity which identifies exposures to risk.

Another factor to be studied prior to undertaking a risk assessment is the existence of any policies, rules, regulations, goals, standard operating procedures or instructions related to security that have been or are presently in force. It is important to review all of this data in order to know what has already been done and to understand the reasons for current security procedures. Perhaps earlier recommendations had been extensively tested and were found to have little or no effect, so that a revision of the policy would be of no benefit. This review should determine what current security-related policies do exist, or if new policies must be developed. The review may also provide guidelines against which to measure the existing security policies and give reasons for updating or creating additional policies.

Risk assessments can be made by security specialists presently employed by the organization or by outside consultants hired for that purpose. There are advantages and disadvantages to both methods. The use of a present employee has the advantage that he will know the physical and organizational structure of the organization, its goals, and its own method of operation. On the other hand, security employees are too often overwhelmed by current crises or interdepartmental rivalries, and have limited authority or status.

One of the advantages in employing an outside consultant to conduct risk assessments is that the consultant has no organizational loyalties to contend with, which helps reduce competitiveness and internal rivalries. An outsider may observe threatening factors that have been accepted as normal conditions, or something that cannot be changed, by a regular security employee. Yet, the outsider would suffer the disadvantage of limited exposure to the organization and its environs, the diverse personalities and attitudes of the employees and management, and the subtle interplay of the formal and informal power structures within the organization. The outsider may attain only a superficial understanding of the situation if the survey is constrained by inadequate time and support factors.

Large organizations with full-time security directors tend to be reluctant to admit the need for assistance from outside, regardless of how limited in experience their own personnel are. Most smaller organizations assume they cannot afford to hire outside consultants, and many feel they can acquire adequate security advice from their insurance agents or local police departments. There are many organizations that feel that an outside consultant would be primarily interested in the sale of security hardware, regardless of its need. For these reasons and others, risk assessments are usually performed internally, utilizing standard form checklists.

Finally, an effective risk assessment can be attained only when there is wholehearted support by both management and operational personnel. Management must provide the authority and financial support necessary for the survey. Still, there will be some resistance to the intrusion of security personnel into areas previously thought to be private domain by certain employees. This problem will be discussed throughout the chapter.

Risk Assessment and Loss Probability

The purpose of a risk assessment is usually stated to be the determination of the vulnerability of a specific organization, facility, or activity to hazards or dangers caused by natural or man-made forces. The person who is selected to conduct the assessment frequently receives no further directions or instructions. He will have to develop a survey plan based on the fact that the total security of an establishment is possible, but that total security may not be cost-effective or supported by the top management. Therefore, priorities must be developed, so that both high- and low-value items receive security commensurate with their value.

Establish Criticality

The first step in a security survey, then, is to determine the value, impact, or cost of any asset, should it be lost as the result of natural or man-made forces. The person conducting the survey will need to ascertain from the operational personnel the identification and relative value of the buildings, equipment and activities in the area to be surveyed, and how important they are to continued productivity for the entire organization. For example, in a factory complex, one of the buildings may house manufacturing processes that require unique, sophisticated equipment. If the operations of the entire factory depend completely upon the product of that special equipment which is either irreplaceable or would take so long to replace that the company could not survive, its protection should be afforded a high priority over the operations that could be performed in other spaces.

Thus, priorities for protection should be assigned so that the areas most critical to the survival of the organization receive the best security, while non-critical areas should be assigned a lesser priority for protection. In the situation where two facilities have equal production processes and output capability, numbers of personnel, and structural size, yet one is self-sufficient in its supply of necessary water or power, whereas the second is supported from external sources, the more self-sufficient facility may be given a higher priority, because it would be affected least by some outside interference.

The same principles for assigning priorities for security apply to all types of facilities. For example, one can assign priorities for security in a small retail store. There the location of the items of the highest value should be ascertained. One problem may be that during the business day the area with the highest value items might be the jewelry or camera department, whereas at night all cash is collected from throughout the store and placed into a safe in the office. In such a situation the high priority area for security changes from day to night.

As noted earlier, the person conducting the survey will need to gain the cooperation of the operating personnel in order to discover the actual processes followed throughout a normal period of operation. When surveying small businesses, it is not unusual for top management personnel to outline the procedures for the handling of cash or high value items, and then to learn from the operational personnel that the cash is hidden in the building overnight to avoid a trip to the bank's night depository. As the surveyor verifies each of the actual procedures undertaken by the operating personnel, a decision must be made whether to assign each one a high or low priority rating.

After a facility has been inspected by the surveyor, and the ratings of priority have been given to the various physical areas or operational phases of the unit, the ratings can be displayed on a chart that will graphically identify the priorities assigned. Table 10-1 is an example of the results of the priority ratings assigned to the major units in a small production facility.

Frequently, situations arise where certain management or operational personnel strongly oppose the priority ratings arrived at by the surveyor. One way to forestall such opposition would be to form a three-person committee with representation by interested and knowledgeable management and employee personnel, as well as the surveyor. Each committee member could assign his own ratings of priority and then discuss with the other two members his reasons for the rating, if all three committee

Table 10-1
Priority assignments.

	Very high Priority (A)	High Priority (B)	Some priority (C)	No priority (D)
Entrance			X	
Visitor's lounge				X
Cafeteria			X	
Employee lounge				X
Personnel office			X	
Executive office		X		
Management office			X	
Auditor's office		X		
Finance office	X			
Shipping and receiving	X			
Production rooms			X	
Warehouse		X		
Inflammable storage		X		
Vehicle maintenance			X	
Security gate and office			X	

members do not agree. Surprisingly, fair ratings often can be attained by averaging the ratings by the three committee members.

One way to make the priority ratings more objective is to determine actual replacement figure costs. Once the costs of replacement, down-time, lost customers and employee diversion or unemployment are tabulated for the various units of a facility, the rating of priorities becomes a simple and objective matter.

Determine Vulnerability

The second criterion of a risk assessment is to determine the degree of vulnerability the facility or activity has to damage or attack by natural or man-made forces. Empirical data has shown that buildings with the larger number of openings to the outside (doors, windows, skylights, etc.) are the object of more man-made attacks than buildings with fewer openings. Similarly, buildings with fences, lights, and guards are victimized less frequently than buildings without such security. The vulnerability to attacks of the lesser protected buildings is higher than better protected buildings. The amount of criminal activity in the surrounding area should also be taken into account.

The vulnerability risk is also heightened by the value and size of items that are maintained in the facility. Large amounts of cash and small but high value items such as jewelry, cameras, or drugs are more frequently the target of theft than are pre-cast concrete sewer pipes, structural steel, or crude oil in 500,000 gallon storage tanks.

The vulnerability of an organization, facility, or activity to natural forces must be analyzed somewhat differently. Again, local records should be reviewed to determine the frequency and severity of earthquakes, floods, or other disastrous natural forces. Still, recent records may not be sufficient to predict the possibility of threats of natural disasters, so state or federal level sources should be queried concerning the actual vulnerability of the local area to such natural phenomena. Further, certain types of physical structures are more susceptible to damage or destruction than others, so assistance should be solicited from engineers to inspect the soundness of buildings included in the survey. Finally, the vulnerability to theft or looting is increased after natural disasters, and this must be taken into account.

There are several ways to determine the vulnerability of a facility or activity to damage or attack. The first is for the experienced surveyor to personally inspect the facility for physical or operational weaknesses. Such an inspection entails more than one visit. The facility should be observed during peak operational periods as well as during closed or slack times. Nighttime visits are important, even if the facility is non-operational. The surveyor should inspect the facilities to ascertain the degree of security and to discover weak points. Chapters 4 through 8 provide an overview of protective devices that are commonly in use. The absence of such security hardware or the inoperability of installed hardware may be a source of vulnerability to the assets of an organization.

Another method to ascertain the vulnerability of a facility is to look at its records of losses, and to determine whether all losses have been properly reported and recorded. A high or obviously increasing rate of losses should alert the surveyor to a vulnerability problem. One must note here that many losses are not reported to higher management because the operational personnel are afraid to do so. Many firms are self-insured and simply absorb losses without any further action. Some businesses fail to report losses because they have been advised that additional claims for losses will cause a rise in their insurance premiums. In cases such as these, there will be few formal records of losses. The surveyor will have to get such information from the employees by routine interview methods.

A third way to determine the vulnerability of a facility is to determine whether the high-value property or items are properly safeguarded from theft by insiders. Frequently the location for high-risk items is chosen without the problem of internal theft in mind, and may be near exits, restrooms, or trash bins. Such locations should be analyzed, and if there is no real need for the high-risk items to be located there, they should be relocated to a less traveled or open area.

Retail stores most frequently locate the cashier's office, camera department and other high-risk areas some distance away from outside doors.

The knowledge that a long route must be traversed in order to escape does have an effect upon the number of transgressions occurring deep within the store.

As the survey is conducted, a chart can be prepared that will illustrate the degree of vulnerability of the various areas. Table 10-2 is an example of the combination of a listing of high value items located in various units and a listing of the losses that have taken place during the past year. Based on Table 10-2, it can be seen that if no increased security measures have been taken since the previous losses, that property in the warehouse and in the shipping and receiving areas are most vulnerable to loss.

Table 10-2
Graphic display of vulnerability to loss, and loss in previous year.

	Cash on hand	Pilferable supplies	High value items	Loss past year
Entrance	—	—	—	—
Visitor's lounge	—	—	—	—
Cafeteria	yes	yes		$300 cash
Employee lounge	—	—	—	—
Personnel office	—	yes	yes	$125 adding machine
Executive office	—	—	—	—
Auditor's office	—	—	yes	—
Finance office	yes	yes	yes	$30 pocket calculator
Shipping/Receiving	—	yes	yes	$2,000 inventory shortage
Production rooms	—	yes	—	—
Warehouse	—	yes	yes	$3,500 inventory shortage
Inflammable storage	—	yes	—	$5 paint
Vehicle maintenance	—	yes	—	$80 carburetor
Security gate and office	—	—	—	—

Probability of Occurrence

The third criterion applicable to a risk assessment is the degree of probability that natural or man-made forces will strike any given organization, facility, or activity. There is a high probability of disastrous storms along the hurricane belt. Blizzards are commonplace in the North and Midwest. Low-lying valleys with rivers or valleys below dams stand a high chance of flash floods during rainy seasons. Earthquakes occur with regularity in the West, and even volcanoes erupt. The degree of probability of such occurrences can be estimated with fair success when based upon past experiences. Attacks by man can likewise be predicted after all the high priority and vulnerability factors have been considered.

Generally, one can observe the rate of past experiences of natural or man-made forces and determine whether there is an increase, decrease, or relatively constant number of events occurring over a period of time.

When coupled with such factors as high priority for security, a high vulnerability risk, and frequent devastating storms, the probability of loss is high. The lack or infrequency of natural disasters, the absence of physical losses, and low impact in event of loss would predict a lower or insignificant loss probability.

The loss probability can likewise be estimated for attacks by man. A chart can be prepared to show the number of known depredations committed for various periods in the past. There is some value to reviewing the loss records for as long ago as ten or more years. Such a review might disclose a favorable or unfavorable record of security, but more importantly the record allows a comparison with the rate of criminal activity during the past year. Increasing or decreasing trends in crime can be determined and may provide top management with the data that will support a request for a stronger security policy or additional funding.

The information collected in Table 10-3 tends to show that in the past year a slightly more than average number of criminal acts had been recorded in one production facility. This should predict an increasing probability of loss and indicate a need to upgrade security policies and provide more or better security measures. There does not appear to be an increase in destructive natural events. However, because there had been an earthquake in the area in the past one hundred years, the probability exists that another earthquake might occur. Otherwise the prediction for the probability for loss by means of natural forces is minimal.

Table 10-3
Probability assessment scale: Frequency of occurrences in the past.

	100 years	50 years	10 years	1 year
Earthquake	1	—	—	—
Tornado or windstorm	unknown	75	15	2
Flood	unknown	30	7	1
Blizzard	unknown	unknown	3	1
Other	—	—	—	—
Burglary	—	—	20	4
Robbery	—	—	1	—
Vandalism	—	—	29	5
Internal theft	—	—	4	3
Fraud	—	—	3	2

Relationship of Security to Liability

A phenomenal outgrowth of the better educated citizen, of the high technology age, and of an expanded business and industrial climate has meant a veritable tidal wave of lawsuits. Every business, every profession, every industry has been affected. Some accounts of this problem report payments in security-related lawsuits have been increasing at a rate of 300 percent annually since 1967. Litigation against businesses in which security services had no direct involvement have increased at a similar rate. A brief look at some of the issues of the lawsuits will provide additional guidelines and impetus for the conduct of security surveys.

Security related lawsuits have affected virtually every type of organization that employs security personnel, including stores, hotels, hospitals and schools. The largest number of payments were for "inadequate security" or a failure to prevent crimes in motel or hotel rooms, hallways or parking lots. The next largest group involved crimes committed by employees, improper security actions against invitees and libel over employee crime. Non-security related lawsuits that could have been prevented were for physical injuries due to faulty or poorly maintained equipment, sidewalks, stairwells or lighting.

There are several factors that seem to be involved in this exceptional growth of litigation. One profession recently claimed that the contingency fee basis and the excessive number of practicing attorneys, coupled with the "deep pockets" of the profession or their insurance companies, made a tempting target. The security field has also seen a large growth in personnel and equipment, which is a very visible target when improper action is taken or the equipment malfunctions. Those businesses who will not or cannot afford to provide the same level of security found in similar businesses can be faulted for not providing a standard level of protection. The courts have become so active in their interpretations of standards of security that it is difficult to find two of their definitions of "adequate security" that agree.

Some of the cue words in the holdings cited by the courts are that security forces and their employers have a "duty to protect," not only employees, but invitees as well, that injurious acts, especially crimes, are "foreseeable," and that there was no "adequate security" provided. This latter aspect includes complaints of inadequate selection, training, or supervision of security. Perhaps greater care should have been paid to these factors in the earlier years. Now security forces and their employers have no choice but to improve their services.

These problems of civil and criminal liability are a new challenge for the physical security survey. The surveyor must now approach each facility with the goal of identifying situations or activities that could potentially lead to lawsuits, as well as identifying the usual risks to lives and property. The inclusion in the survey report of obvious safety hazards and violations of OSHA regulations can be an immeasurable cost savings to the employing firm. These prevention tactics cost much less than the expense of lengthy litigation.

Designing Security Surveys

The type of security survey to be conducted of any organization, facility or activity is dictated by the purpose of the survey and by the kind of facility to be surveyed. As stated earlier, surveys can be conducted to determine the relative risks of losses from natural and man-made forces, but not all the information gathered by one type of survey will be of help in recommending protective services for the other. Therefore, most surveys will have to be carefully designed to incorporate the specific goals and the unique physical and personnel structure of each organization.

Surveys generally specify one or more of the following listed objectives:

a) Determine the existent vulnerabilities to injury, death, damage or destruction by natural causes.
b) Determine the existent vulnerabilities of corporate assets due to criminal activity from outside the organization.
c) Determine the existent vulnerabilities of corporate assets due to criminal activity from within the organization.
d) Determine the existent conditions of physical security of corporate property.
e) Identify physical hazards, operational procedures or personnel activities that could lead to legal liability for damages or injuries to employees, patrons or the general public.
f) Measure the effectiveness of the current protection policies and standards.
g) Measure the conformity of the employees to the published security standards.
h) Audit the accounts and procedures of the firm to detect policy violations, fraud and use of improper procedures.
i) Inspect the conditions and procedures that cause the problems of inventory shortages, cash or property losses, vandalism, or other unexplained crime within the plant.
j) Investigate the economic-sociological-political conditions in the community to predict outside activities that could be adverse to the well-being or survival of the company.

A cursory glance at these objectives will reveal that the requested surveys may be instigated as a reaction to recurring problems within the organization, or simply as a result of a regularly scheduled program to assess conditions before they become risks or losses.

In practice, there are some security agencies that make a distinction between physical security surveys and crime prevention surveys. The rationale given is that although physical security measures do prevent crime, they are oriented more toward the security of property and facilities, whereas crime prevention measures encompass the deterrence of criminal activity regardless of the extent or availability of physical safeguards.

One operative definition of physical security is: "that part of security concerned with the physical measures designed to safeguard personnel, to prevent unauthorized access to equipment, facilities, material and documents, and to safeguard them against espionage, sabotage, damage and theft." The physical measures cited in the definition are those described in Chapters 4 through 9 concerning perimeter security, lighting, locks, intrusion detection systems, storage containers, and security personnel. A review of those chapters will provide the person who conducts a physical security survey with sufficient guidelines to identify inadequately safeguarded property. In other terms, a physical security survey is directed toward a detailed examination and to specific recommendations provided for the application of physical measures to prevent the opportunity to commit a crime.

The foregoing definition does not conflict with a currently popular definition for crime prevention, adapted from the British Home Office Crime Prevention Program: "The anticipation, recognition, and appraisal of a crime risk and the initiation of action to remove or reduce it." The absence of mention of physical measures has perhaps influenced some practitioners into believing this definition is incompatible with physical security.

Survey Checklists

Despite the proliferation of survey checklists throughout the security industry, few of them are totally compatible with all the survey objectives or are applicable to all facilities. Yet they do provide a guideline to the surveyor, so that the most obvious conditions are not overlooked.

There are advantages and disadvantages to the use of checklists. They do serve as a reminder that specific subjects or areas should be inspected. Checklists can be devised to be used as an outlined draft of the final report. The checklists can direct the examination of the facility from the exterior to the interior, or from the general to the specific features that must be observed and reported. The use of a checklist can also help other surveyors to continue the survey should the initial surveyor have to leave the assignment.

On the other hand, many surveyors do no more than what is prescribed by the checklist, thereby limiting their own contribution. The checklist may have been adapted from another facility or agency and not be totally applicable to the surveyed facility. Surveyors can become accustomed to checking off items rather than describing the situation by their own terminology, which may be more precise. Finally, checklists tend to allow the surveyors to make a more cursory inspection, whereas the narrative report would require a more detailed examination to properly describe the item reported.

A sample of survey checklists is included in Appendix B.

There are no hard and fast rules on how to conduct a survey. Some have been performed as hasty walk-throughs with a single page checklist, while others involve lengthy preparation and visits for several weeks. The more experienced surveyor may be able to complete a survey in a fraction of the time it would take a novice. In any case, the following design of survey activities is offered as a guide to follow if there are no further instructions available from the requesting office.

Preliminary Activities

Written authority to conduct the survey should be obtained. This authority should clearly outline the goals and objectives of the survey and establish time, support and availability of other administrative facilities for the use of the surveyor.

The historical, geographical and other background data of the facility and its environs should be reviewed. The local media and public records should be checked to establish the relationships and attitudes of the facility and the local community. Local crime rates should be compared with the rate of crime occurring in the facility.

A review of the written policies, rules, regulations, standard operating procedures or other instructions relating to security should be made.

Maps, floor plans or architectural drawings should be obtained to determine the main characteristics, especially the locations of doors and windows.

A review of any previous surveys that have been reported should be made.

A checklist containing questions that must be answered to satisfy the initial purpose of the survey should be assembled.

Performing the Survey

The surveyor should review, with the management personnel and supervisors of the organization, the facilities or activities to be surveyed and the purposes of the survey. This review should engender their cooperation and solicit their continued contributions to the survey.

An orientation tour should be made to establish the limits of the facility. The area supervisors should be solicited for their concept of the corporate security policy and standards. They should also demonstrate all security procedures applicable to their area of jurisdiction.

Arrangements should be made to review these procedures with supervisors on all shifts.

The facility should be visited during peak and low operating periods to personally observe the personnel, procedures and physical measures, including the perimeter barriers, the lighting, locking devices, intrusion detection systems, storage containers and the security personnel.

The survey checklist should be expanded wherever there is a need to do so. Further, narrative descriptions of any features of the security program should be recorded if necessary. Photographs should be taken of objects or situations that are difficult to describe.

Coordinating Interviews

There should be a review with the local supervisors of all the deficiencies that were noted so that immediate corrective action can be taken. In some security programs, the local supervisors must prepare a written list of corrective actions taken shortly after this review.

On occasion, the local supervisors will quarrel with the survey findings. For this reason and for the sake of fairness, the surveyor must be careful to point out only genuine deficiencies in security, and not employ harassing tactics or exaggerate in order to emphasize his own importance.

The Survey Report

Survey reports must follow the requirements of the requestor. Some reports are merely cover letters attached to the checklists. It may be true that the requestor need not be provided with details beyond a brief list of deficiencies and recommendations. The extent of the details of the survey report should be prescribed by the original request.

Survey reports may spell out a complete narrative description of the physical plant and the prescribed security policies or standards, so that the reader may make a direct comparison of the report with the facility surveyed.

When there is a notation in the report that there is a deficiency because an existent security measure does not meet the organizational standard, it would be redundant to add that the deficiency should be removed, but recommendations should be fully described.

The Follow-Up Survey

A follow-up is performed to ensure that deficiencies have been corrected or are actively and consistently repaired.

It should be scheduled after a sufficient period of time has passed in which corrective action can be taken to eliminate the deficiencies. This period may be 30, 60 or 90 days. Any period longer than that may leave security risks uncorrected for too long a time. The follow-up survey reviews only those areas with deficiencies.

Any evidence that the deficiencies and recommendations have been ignored must be taken to the top management level of the organization.

Frequency of Surveys

There are no strict rules that prescribe how often surveys should be conducted in a specific facility. Corporate-level security policy should provide some flexible guidelines for the security surveyor to follow. The following listed conditions may help to determine the appropriate frequency of surveys.

In the case of a new organization, facility, or activity, a thorough survey should be made to ensure that there is a complete record of the security measures and any security deficiencies. The frequency of subsequent surveys should consider the extent of the risks of vulnerability, priority (criticality), and probability. If the facility is in a high risk category and depredations against corporate property are constant, surveys should be scheduled at greater frequencies, perhaps as often as quarterly or semi-annually. When the facility appears to have a low risk rating, annual surveys should be considered.

One method to provide the impression that security has a high priority is to conduct inspections of units of the larger organization throughout the year after the initial survey. In this manner, smaller inspections or audits of the physical security measures or of operating procedures, alternated throughout the organization, may provide a habit of security consciousness among employees. The conducting of organization-wide surveys would not need to be more frequent than annually. The smaller units inspected may be selected on a rotational or random basis, or they may be selected based upon the number of security problems that surface after the initial survey.

Frequency of security surveys may also be increased after serious or repeated depredations against the firm. However, the actual point of attack may not be the weak link in security, so any investigation of the security problem should extend beyond the obvious scene. In the case of repeated attacks, the survey frequency should be increased until a way has been found to solve or reduce the problem.

New surveys should be considered every time the facility reorganizes its physical structures or personnel, or after the acquisition of new buildings or expensive equipment. The risk probability, vulnerabilities and security priorities will no doubt change under those conditions. Extensive changes of property or equipment indicate a need for all-inclusive surveys, but minor changes or mere additions of new equipment in already high security areas may need only a less inclusive update survey.

These suggestions do not include every reason why a new or additional survey should be made. It is the responsibility of the security supervisors and top management to promptly evaluate any change that may increase risk to the organization's well-being or survival.

Cost-Benefit Analysis

Not all commercial or industrial firms or institutions have regularly organized security plans, programs, or security personnel. About 58 percent of the ten million businesses in this country gross less than $100,000 annually. Taken together, these smaller firms suffer larger losses from criminal action and employ a disproportionately smaller number of security measures to protect their assets. Additionally, the smaller firms suffer a higher rate of bankruptcies, and not a few of the bankruptcies are caused by shoplifting and employee theft. Small business owners complain that they cannot afford security, but in reality they cannot afford *not* to have security.

At first glance, an outsider might conjecture that security personnel are only window dressing, and do not contribute productively to business firms. That line of reasoning is fallacious at best. Even in the old days, when every factory had its own guard to watch the buildings for security purposes, the guards were additionally productive by oiling machinery, emptying trash cans and providing other non-security services. They may have earned more of their pay by their other services, unless they were lucky enough to prevent a fire, thwart a burglary, or perform some life-saving act for an employee. But those kinds of events were rare and are still relatively rare today.

The most difficult task of the modern security manager is to justify the expenses of security. The security program is still largely seen as a group of little old men making certain that the coffee pots are unplugged and that some worker has not thrown a smoldering cigarette butt into a paper-filled trash can.

The security manager can change that old image of the security department by preparing a cost-benefit analysis for his unit. A cost-benefit analysis is a direct comparison of the costs of the operation of the security unit and all security measures with the amount of corporate property saved or recovered and elimination or reduction of losses caused by injuries and lost production time. Despite the claim that it is impossible to compute the savings incurred through the crime prevented by the security officer's presence, there are methods for computing the relative values of security measures.

The first rule of a cost-benefit accounting system is to not recommend a security measure that is not cost-effective. That is to say that one should not spend $10,000 to protect a $5.00 object. Nor should one erect a $10,000 fence to secure industrial salvage worth $100. But a $10,000 fence would be cost-effective if it were built to protect something valued several times greater than the fence or eliminate a guard position that is more costly.

Extending that idea, the security manager should be able to show over a period of time that the services of his department at a cost of X dollars have effectively decreased losses from X dollars in year so-and-so to Z dollars this past year. In other instances, the security manager may be able to show that although the remainder of the community had a crime rate increase, his jurisdiction actually had a decrease.

Although a direct dollar loss or gain cannot always be seen in such a direct comparison, some evaluations may be estimated for illustration purposes.

Every structure and its contents has a dollar value assigned by the organization. Estimates of the cost of fire, vandalism, or theft can be made and projected showing the probable rate of occurrence if security is not effective.

Estimates of lost time due to theft of equipment or materials would likewise present a good argument that security is cost effective in its preventive measures.

There are often some reductions in insurance premium rates if alarm systems, fire sprinklers, or guards are employed, which contributes to a cost saving.

If the security personnel become more adept at their job and actually improve their apprehension and recovery rate, the total values of the property saved and recovered and decreased expenses may show a surprisingly satisfactory cost-benefit relationship.

Ultimately, the security manager may investigate the possibility of replacing some of the security personnel with electronic devices that would cost a fraction of the guards' annual salary. Not only would the operating budget be lowered by such mechanisms, but management may become so appreciative with the cost-benefit results that the money saved through the listed methods might be returned to the security unit's budget as an increase the following year.

Selling the Security Plan

There is no single method for selling a security plan to management. The best way to sell a security plan is to apply some basic logical principles to the preparation of the security plan, so that management will readily recognize that the proper work has been done.

The traditional steps in planning should be followed when a security measure is proposed: (1) recognize a need, (2) state the objectives, (3) gather the relevant data, (4) develop alternatives, (5) prepare a course of action, (6) analyze the capabilities, (7) review the plan, (8) present the plan to management.

The starting point of any plan is when one recognizes that a security problem exists. A problem may be indicated by a series of losses or cases of vandalism. It should be ascertained that the losses are not coincidental, and that there really is a problem. The second step would be to state the objective of the plan. The objective may be: investigate the causes of the recent losses of lead pipe and devise measures to prevent future thefts effective within 30 days.

In order to determine how the lead pipes were lost, records and people must be used to verify the possession and subsequent loss of pipe. These sources of information might lead one to identify the perpetrator, but at the very least, they will provide data for the next step, that of developing alternatives. Various methods of securing the pipe in the future should be listed, such as (a) securing in a locked shed, (b) placing in a locked, fenced area, (c) posting a guard.

After analyzing the alternative courses of action (and keeping in mind the cost and benefits), the one most favorable to the firm should be selected for presentation to management. If there are presently insufficient guards to post one at the pipe storage area, and the budget precludes additional hiring, another alternative should be selected. This final arrangement should be reviewed as to its feasibility and prepared for presentation to the management person responsible for the final decision.

The presentation should also be made with consideration of the organization's financial situation at the time of the proposal. If the firm is cutting back its budget, any savings suggested by the plan should be highlighted. If the organization is in an expansion phase, the presentation should emphasize how the proposal fits into such a plan. The security manager should be able to dovetail any proposal into the firm's present plans, otherwise there may be a summary rejection, because the plan is out of step.

Most proposals for changes or additions to programs must be presented to management in such a fashion as will favorably attract their attention. Clean, well-prepared reports, audio-visual aids, large charts, etc., will serve to get across the message that effort was expended to present this case. Nothing should be left to chance. The more that is known about the plan that is to be sold to management, the greater will be the confidence of the presentor. And the greater his confidence, the easier it will be to convince management that this plan will work.

There is no guarantee that all proposals will be accepted by management, but completion of these steps will provide for a better presentation.

Maintaining Corporate Support

After the new security plan has been sold to management, the security manager will have the responsibility for implementing the plan and maintaining management's continued support for it. Without both moral and financial support, most programs seldom last very long.

One of the most effective ways for the security director to maintain the interest and support of management is to become a more involved member of management. There are many ways he can become more active and accepted, including the ready acceptance of added responsibilities, the participation in the less popular programs, and an attitude and demeanor that will set an example of honesty, seriousness and dependability.

The security director, by virtue of his position, must be the leader, not only in his own specialty, but among other activities of the organization. Volunteering to assist in fund drives, setting the example by quickly responding to those in need, and constantly serving selflessly will engender admiration and support from all levels of personnel. In that respect, management would be hard pressed not to support their most proficient manager.

Discussion Questions

1. List the steps necessary before action can be undertaken to minimize losses by injury, death or damage within an organization.
2. Why is it necessary for a surveyor to become familiar with company policies, regulations, goals, and standard operating procedures, before a risk assessment is initiated?
3. List some advantages and some disadvantages in having risk assessments conducted by persons from outside the organization.
4. Not all personnel, activities, or facilities have equal needs for protection. Identify the criteria for establishing priorities for security.
5. Discuss the methods used to assess the vulnerability of a facility.
6. Identify several purposes of surveys of organizations.
7. Discuss the advantages and the disadvantages of survey checklists.
8. Outline the major steps of a risk assessment survey.
9. List the traditional steps of planning that can be utilized to propose security measures to management.

References

Army, Department of the. *FM 19-20 Military Police Criminal Investigations.* October 1971.

__________. *FM 19-30 Physical Security.* November 1971.

Cole, Richard B. *Protection Management & Crime Prevention.* Cincinnati: The W.H. Anderson Company. 1974.

Colling, Russell L. *Hospital Security*. Los Angeles, Calif.: Security World Publishing Co., Inc. 1976.

Comer, Michael J. *Corporate Fraud*. London: McGraw-Hill Book Company (UK) Ltd. 1977.

Ehrstine, B.I. and J.A. Mack. *Profitability Through Loss Control*. Cincinnati: Anderson Publishing Co. 1977.

German, Donald R. and Joan W. German. *Bank Employee's Security Handbook*. Boston: Warren, Gorham and Lamont. 1982.

Healy, Richard J. and Timothy J. Walsh. *Industrial Security Management*. American Management Association, Inc. 1971.

Hemphill, Jr., Charles F. *Management's Role in Loss Prevention*. AMACOM, a division of American Management Association. 1976.

Kingsbury, Arthur A. *Introduction to Security and Crime Prevention Surveys*. Springfield: Charles C. Thomas. 1973.

Mandelbaum, Albert J. *Fundamentals of Protective Systems*. Springfield, IL: Charles C. Thomas. 1973.

Moore, Kenneth C. *Airport, Aircraft and Airline Security*. Los Angeles: Security World Pub. Co. 1976.

San Luis, Ed. *Office and Office Building Security*. Los Angeles: Security World Pub. Co. 1973.

Ursic, Henry S. and Leroy E. Pagano. *Security Management Systems*. Springfield, IL: Charles C. Thomas. 1974.

Weber, Thad L. *Pharmacy Security Manual*. Philadelphia: Smith, Kline and French Laboratories. 1976.

Chapter 11
Internal Threat and Crime

The impact of internal thefts (and other problems initiated by employees) on business and governmental organizations has reached alarming proportions. Fraud and embezzlement are two major crime categories which account for major losses in various enterprises. These and other internal thefts may be written about, noted or discussed in a number of ways in order to try to disguise what has occurred with property that belonged to the organization and is now missing. Some companies simply refer to internal losses as "shortages" rather than try to explain how or why their employees may be involved in stealing company property.

According to the National Business Crime Prevention Network, an Atlanta-based business abuse consulting firm, American businesses will lose well in excess of $40 billion every year.[1] A staggering figure, yet, it is only the tip of the proverbial iceberg as it represents only the tangible and measurable losses. Added to the costs of security services and hardware, investigative services, associated legal expenses and the negative impact on human resources and productivity, the overall sum may be ten times the original figure, or nearly $400 billion annually.[2]

In 1984, internal thefts, totalling $382 million, accounted for 86% of all losses to financial institutions, according to FBI statistics. The average inside bank fraud or embezzlement amounted to $51,738, more than eight times greater than the average outside burglary, larceny or robbery.[3] Thus, about $8 was taken by an employee for every $1 taken by an outsider.

Various studies have indicated that one in three to four employees steal from their employer. Given a total employment population of around 110 million people one can draw the supposition that some 27.5 million to 37 million workers have stolen money, goods, merchandise, etc., from their employers. Obviously, such a supposition is not based on a factual data source, however, it becomes much easier to see how many insignificant acts of employee theft by millions of employees, coupled with large scale incidents of employee fraud and embezzlement by a much smaller segment of the workforce, can amount to annual losses in the billions of dollars.

The above statistics indicate that internal theft is a problem of epidemic proportion, and if left unchecked, marginal enterprises cannot exist when direct and indirect losses begin to approach their after-tax gains. Richard Hefter refers to employee theft as *The Crippling Crime*, detrimental and often devastating to four distinctive and important areas of business concern:

1) *Insurance costs raise.* Losses have caused fidelity underwriters to drastically increase their premiums to insure against employee dishonesty. Costs in some areas have reached epidemic stages.
2) *Increased funds invested to prevent losses.* Security protection that will be needed will increase especially in the areas of deterrent devices and methods investigations and prosecutions.
3) *Increased prices for goods and services.* Increased prices due to internal thefts have caused many retailers to lose their competitive position. In fact, the need to increase prices and the loss of customers as a result of the increases has caused many businesses to go bankrupt.
4) *Employee productivity and morale decline.* Employees involved in thefts or associated with employees involved in thefts will have a decline in both productivity and morale. Quality of service declines, if the business is service-related, and errors occur if business is producing a product. In either case, the reputation of the company declines.[4]

Employee Embezzlement

One of the major internal theft categories is embezzlement. Embezzlement is defined as the "fraudulent appropriation for his own use or benefit of property or money entrusted to him by another, by a clerk, agent, trustee, public officer, or other person acting in a fiduciary character."[5] Embezzlement is common-law larceny extended to cover cases which do not include a trespass, or cases where an individual may abstract or misapply funds or property of another person or entity.

The statutes defining the crime of embezzlement do not make a distinction between a cash loss and a property or merchandise loss. However, according to several insurance groups, losses from embezzlement of cash reserves are seven to ten times higher than losses from property or merchandise. Even though losses from cash reserves far outweigh the losses from merchandise, merchandise loss is still an embezzlement problem.

In order to properly ensure that the risks from embezzlement are reduced, an employer must be familiar with the human and physical elements that are generally present in an embezzlement. These elements and their components are need, rationalization, and opportunity.

1. *Need*—A physical condition (sickness) may be the cause of unusual debts and may cause an employee to think seriously about how to convert merchandise or money to his/her own use. Bad health coupled with the stressful necessity of having to pay a large debt may trigger an employee into committing an embezzlement act. Other personal acts of an employee, such as overcharging on one or more credit cards or by financing several items over a period of time with different due dates, may cause an employee to initiate a theft. Even though personal gain is far more common than the need for revenge or philanthropy, they certainly cannot be disregarded when one considers the component parts of the element of need. Several years ago, in a midwestern state, a savings and loan company discovered that a very trusted female employee had some unusual philanthropic endeavors. When questioned, she admitted to the acts of embezzlement and stated that she gave most of the embezzled funds to local charitable organizations.

Personal financial gain represents the major reason why persons embezzle from their employer. The psychological needs of an individual are difficult to discover and may change over a period of time, so it is practically impossible to identify an unmet need at any given time and prevent an act of embezzlement.

2. *Rationalization*—This represents the state of mind of an individual before and after a theft has occurred. Rationalization is the psychological state which provides explanations or excuses for one's acts, usually without the individual being aware that these are not the real motives. These false motives are psychological defense mechanisms which allow an individual to face his actions on a day-to-day basis. These rationalizations are usually categorized in four ways: (1) "borrowing, not stealing," (2) lack of moral restraint, (3) moral right, and (4) reward within the work group. The "borrowing, not stealing" rationalization occurs when the employee thinks that he will return the money. Indeed, the employee may return the money in the beginning, but the amount taken over a longer period of time is usually too large to pay back at any one time. Time and need will work against the embezzler. The lack of moral restraint rationalization happens when an employee observes other employees involved in larcenous acts, and after assessing the risk, tells himself that it's all right if he does it too. The moral right rationalization occurs when an employee believes he deserves more pay and is taking only what is rightfully his. The last rationalization category involves the individual who sees other workers stealing and not getting caught. This worker rationalizes that it must be all right because no one is notifying the supervisor, and the thieves are not rejected socially by the work group.

3. *Opportunity*—Certain conditions must exist for the act of embezzlement to occur. Opportunity is usually afforded by management through either omission of controls or the inadequacy of existing controls. The control of opportunity is the key to controlling the elements of need and rationalization. The lack of opportunity certainly does not eliminate the need to embezzle, but it does prevent the act from happening.

An employer should be observant and watch for the following behavioral conditions in an employee who has been associated with embezzlers in the past. First, the employee makes sudden changes in his/her spending habits which are obviously above the employee's regular standard of living. These changes could be a new house, new car, new boat, expensive clothes or expensive parties or trips. Also, an employer should be aware of an employee who has had a major illness or a member of the immediate family has had such an illness and a large indebtedness occurred. Second, an employer should observe an employee who has an apparent devotion to his work. This should be especially true of an especially trusted employee who may handle large amounts of money or who may be approving purchases or vendor contracts. In the case of a purchasing agent, funds or services belonging to the company may be given to the employee without the company knowing about the transaction. Third, an employee in a fiduciary position who strongly objects to procedural changes or closer supervision should be audited or observed. This resistance may be fear of the discovery of an illegal act.

Employee Fraud

Fraud is an essential element of various statutory offenses involving theft, misappropriation, and inventory shrinkage. These offenses, while having the same or similar elements, are given different names in various jurisdictions and are scattered throughout the statute books in the theft offenses against property and commercial practice chapters.

Fraud is an intentional perversion of truth for the purpose of inducing another in reliance upon it to part with some valuable thing belonging to him or to surrender a legal right.[6] Fraud includes all acts, omissions, and concealments which involve a break of legal or equitable trust or confidence justly reposed and are injurious to another or by which an undue or unconscionable advantage is taken of another.

Embezzlement (sometimes called fraudulent conversion) may be defined as the unlawful taking of the property of another by one to whom the property has been entrusted. This contrasts with theft, which is simply the unlawful taking of property by someone to whom it had not been entrusted.

A common form of embezzlement involves money being paid by a third person to a clerk for the clerk's employer and the clerk appropriates it before it is put in the cash register. Another variation is where goods are delivered by the shopkeeper to his driver for delivering and the driver appropriates the goods to his own use. Embezzlement requires a relationship of trust and confidence between the person taking the property and the owner, combined with fraudulent (unlawful) intent.

According to Bologna, *The Fine Act of Fraud Auditing*, employee fraud, theft, and embezzlement are more prevalent in some organizations than in others. He states that employee crime is a product of, and is distinguishable by, certain environmental and cultural contrasts that impact organizational vulnerability (figure 11-1).

Employee Pilferage

Pilferage is the stealing of property of another in small amounts over a period of time. The dollar value of items taken is small in each incident, but when taken together, incidents of pilferage can amount to enormous dollar losses to the victim. The pervasiveness of employee theft has not been established, as no one actually knows what portion of inventory shrinkage can be attributed to dishonest employees. Some writers estimate that anywhere from five percent to seventy-five percent of employees steal. And, since only a small percentage of employees would have the opportunity to embezzle or be involved in a fraudulent action in the work place, the greater number of employees who would steal would do so by committing simple acts of pilferage.

While pilferage too is a form of embezzlement (since it occurs while in a position of trust and/or responsibility), it is not usually seen as a means of substantive financial gain. Instead, employees are often impervious to the dollar value of what they are taking and have little difficulty in rationalizing their dishonest act.

Most acts of employee pilferage are likely to be very simple, as little planning and preparation is needed for the actual theft of company property. This assumption can be based on the following:

1) Worker theft is generally presumed to be largely undetected, unreported, unrecorded, and unprosecuted—with the result that the vast majority of perpetrators are unknown and go unpunished;
2) The actual act of pilferage by employees is usually accomplished by simply walking out of the workplace during the lunch hour, at closing time or after hours with the item in their possession either concealed in clothing, packages, lunchboxes, etc., or taken when concealment is not even necessary.

While the taking of a pencil, a pad of paper, or nuts and bolts may be considered insignificant by both the employer and the employee, one has but to multiply these individual incidents by thousands of occurrences

Figure 11-1
The Fraud Environment[7]

High Fraud Potential

Management Style
- a. autocratic

Management Orientation
- a. low trust
- b. X theory
- c. power driven

Distribution of Authority
- a. centralized, reserved by top management

Planning
- a. centralized
- b. short range

Performance
- a. measured quantitatively and on a short-term basis

Profit Focused
Managed by Crisis
Reporting by Routine
Rigid Rules Strongly Policed

Primary Management Concerns
- a. preservation of capital
- b. profit maximization

Reward System
- a. punitive
- b. penurious
- c. politically administered

Feedback on Performance
- a. critical
- b. negative

Interaction Mode
- a. issues and personal differences are skirted or repressed

Payoffs for Good Behavior
- a. mainly monetary

Business Ethics
- a. ambivalent; rides the tides

Internal Relationships
- a. highly competetive, hostile

Values and Beliefs
- a. economic, political, self-centered

Success Formula
- a. works harder

Biggest Human Resource Problems
- a. burnout
- b. high turnover
- c. grievances
- d. absenteeism

Low Fraud Potential

Management Style
- a. participative

Management Orientation
- a. high trust
- b. Y theory
- c. achievement driven

Distribution of Authority
- a. decentralized, delegated to all levels

Planning
- a. decentralized
- b. long range

Performance
- a. measured both quantitatively, and qualitatively and on a long-term basis

Customer Focused
Managed by Objectives
Reporting by Exception
Reasonable Rules Fairly Enforced

Primary Management Concerns
- a. profit organization
- b. human, then capital and technological asset utilization
- c. fairly administered

Reward System
- a. reinforcing
- b. generous

Feedback on Performance
- a. positive
- b. supportive

Interaction Mode
- a. issues and personal differences are confronted and addressed openly

Payoffs for Good Behavior
- a. recognition, promotion, added responsibility, choice assignments, plus money

Business Ethics
- a. clearly defined and regularly followed

Internal Relationships
- a. friendly, competitive, supportive

Values and Beliefs
- a. social, spiritual, group-centered

Success Formula
- a. works smarter

High Fraud Potential—continued

Company Loyalty
- a. low

Major Financial Concern
- a. cash flow shortage

Growth Pattern
- a. sporadic

Relationship with Competitors
- a. hostile

Innovativeness
- a. copy cat, reactive

CEO Characteristics
- a. swinger, bragart, self-interested, driver, insensitive to people, feared, insecure, gambler, inpulsive, tight-fisted, number- and things- oriented, profit seeker, vain, bombastic, highly emotional, partial, pretends to be more than he is

Management Structure, Systems and Controls
- a. bureaucratic
- b. regimented
- c. inflexible
- d. imposed controls
- e. many-tiered structure, vertical
- f. everything documented; a rule for everything

Internal Communication
- a. formal, written, stiff, pompous, ambiguous, CYA

Peer Relationships
- a. hostile, agressive rivalrous

Low Fraud Potential—continued

Biggest Human Resource Problem
- a. not enough promotional opportunities for all the talent

Company Loyalty
- a. high

Major Financial Concern
- a. opportunities for new investments

Growth Pattern
- a. consistent

Relationship with Competitors
- a. professional

Innovativeness
- a. leader, proactive

CEO Characteristics
- a. professional; decisive; fast-paced; friendly; respected by peers; secure; risk taker; thoughtful; generous with person, time, and money; people-, products-, and markets-oriented; builder; helper; self-confident; composed; calm; deliberate; even disposition; fair; knows who he is, what he is and where he is going

Management Structure, Systems and Controls
- a. collegial
- b. systematic
- c. open to change
- d. self-controlled
- e. flat structure, horizontal
- f. documentation is adequate but not burdensome; some discretion is afforded

Internal Communication
- a. informal, oral, clear, friendly, open candid

Peer Relationships
- a. cooperative, friendly, trusting

committed by a multitude of employees every day of the year. The kinds of items and materials that employees steal from their employers include a variety of both finished and unfinished products, office and maintenance equipment and, of course, money.

Numerous studies have shown that people are more likely to steal from businesses and organizations than from other people. There is just now a strong indication that employers are beginning to realize that worker theft is a real problem, and steps are being taken to deter, prevent and detect its occurrence.

To reduce losses through pilferage, many employers use in-house security, contract guard services, and various types of surveillance equipment. Other measures are those that can be designed to monitor or restrict the movement of company personnel in an effort to limit access to certain areas or operations of the facility, thereby denying, or at least making it more difficult to have, the opportunity to steal. This can be accomplished through closer supervision, enforcement of procedural and access controls, increased use of security hardware devices, and increased emphasis of security and its importance at all levels of the work force.

The following discussion of theft control procedures points out some of the basic methods and means that are utilized to control employee theft.

Theft Control Strategies

Most case studies and literature on internal thefts indicate that a positive step which must be taken by an employer is to make all employees aware that internal thefts are a serious problem and that the employer, in the interest of his/her business and its honest employees, must take some action on a regular basis to thwart dishonest acts by employees. The employer should develop and implement at least the following four strategies for reducing and controlling internal theft losses: (1) screening of applicants, (2) procedures or devices which make theft more difficult or apprehension easier, (3) improvement in employee satisfaction, and (4) the policy and process of apprehension and prosecution.

1. *Screening of applicants.* Careful screening of applicants has long been accepted as a way of reducing problems with an employee. As an employer you should investigate the following areas: prior employment, credit references and personal references. Many employers fail to do an adequate job of screening applicants either because of the time involved or because they do not know what they can or cannot ask an applicant. An employer can:

a) Request an applicant to write his/her name and address on an application.

b) Ask an applicant if a complaint has been placed against him/her or if he/she has been indicted or convicted of a crime and under what name.

c) Ask an applicant's age (only if the information is an occupational qualification).
d) Explain to an applicant the hours and days he/she will be required to work.
e) Ask if applicant is a U.S. citizen, or if he/she has the intent to become one.
f) Ask about schooling, both academic and vocational.
g) Inquire about relevant work experience.
h) Inquire into his/her character and background.
i) Ask for name, address, and relationship of person to be notified in case of an accident or emergency.
j) Inquire into applicant's military experience in the U.S. Armed Forces. After hiring, ask to see discharge papers.
k) Ask an applicant about memberships in organizations which do not disclose race, religion, or national origin.
l) Ask an applicant if he/she belongs to an organization advocating the overthrow of the U.S. government.
m) Ask the sex of an applicant only where it constitutes a qualification for the job.

The aforementioned list contains major areas of the applicant screening process which an employer should know and implement. Certain prohibitions are included in the three major laws which govern the applicant screening process: the Civil Rights Act of 1964, the Privacy Act of 1974, and the Consumer Protection Act of 1976. Just as important as the employer's right to know what to ask is the employer's knowledge of what not to ask during the applicant screening process. Under these three Acts a prospective employer may not:

a) Ask an applicant whose name has been changed to disclose the original name.
b) Inquire as to the birthplace of an applicant or applicant's family if outside the U.S.
c) Require an applicant to produce discharge papers from the U.S. Armed Forces (before employment).
d) Inquire into foreign military experience.
e) Ask an applicant's age when it is not an occupational qualification or is not needed for state or Federal minimum age laws.
f) Ask about an applicant's race or color.
g) Require an applicant to provide a photograph with the application.

h) Ask an applicant to disclose membership in organizations which disclose race, religion, or national origin.
i) Ask a male applicant to provide the maiden name of his spouse or his mother.
j) Ask the place of residence of an applicant's spouse, parents or relatives.
k) Inquire whether an applicant's spouse or parents are naturalized or native citizens.
l) Ask an applicant his religion.

More specifically the Consumer Protection Act prohibits requiring information in the following major categories:

a) Records of arrests, indictments, or conviction of crimes where the disposition of the case, release or parole has been more than seven years prior to the date of application.
b) Any bankruptcies which have been more than fourteen years before the application.
c) Any paid tax liens, legal suits or judgments, or other such information which has a harmful effect.

The infrequent use of polygraph tests, the Psychological Stress Evaluation (PSE), the Reid Report and the Reid Survey, although more controversial, have less overall impact on the applicant screening process than does the background check. The polygraph, although not a completely accepted instrument, certainly does not have the same degree of controversy associated with it as does the PSE. The PSE, a technique by which voice patterns are analyzed to detect lying, can be done either overtly or covertly. Covertly, the PSE can be done over a telephone without the knowledge and consent of the speaker. The fact that it can be done covertly is a major point of controversy. The Reid Report and the Reid Survey are two similar versions of a paper and pencil honesty test which were developed and are marketed by the principal author, John E. Reid and Associates. The Reid Report is designed to be used in the applicant screening process, whereas the Reid Survey is designed to be given to persons who are presently employed by an organization.

Many states have statutory provisions regulating the use of the polygraph, the PSE and other such testing procedures. The restrictions imposed by the states have not been included in the above recommendations for a potential employer. It is recommended that someone in personnel review the statutes in their state and possibly the employee contract for prohibitions before embarking on any of the above-mentioned testing procedures.

2. *Procedures and/or devices which make theft more difficult or apprehension easier.* The reader is often made to believe that the problem of employee thefts can be solved by either installing new and intriguing detector devices or modifying or adding to the anti-theft procedures of the organization. Unfortunately, the theft problem cannot be solved by a simple set of procedures or detector devices. The problem should be approached by utilizing both procedures and detectors. It is recommended that the *procedural* approach cover the following:

1) the organization should increase control over company assets and property during the workflow and work assignment,
2) the organization should reduce the opportunity for theft by decreasing the ways that an employee can manipulate records to cover up a theft,
3) the organization should have up-to-date inventory records available and utilize them on a regular basis, and
4) the organization should require that the employees be available for searches of their person and property as they enter and leave the premises.

For example, in an industrial situation, the equipment and merchandise open and available to the employee at any time should have a relationship to the amount of inventory at a work station needed to do a job as well as to the efficiency and effectiveness of getting the parts to the work station. Completed merchandise should be inventoried and placed in a storage area as soon as possible after the assemblage has been made. A direct flow of parts from the stockroom to the work station with no intermediate stops will greatly facilitate control of those parts. Also, for maximum inventory control, the finished product, after it has been completed, should be taken directly from the work station to the storage room.

Allowances will have to be made for breakage, unusable parts and other common reasons for parts loss; however, one would not expect differences between the number of completed products and the number available for shipment to a customer.

The organization should require that employees be available, before entering or leaving the premises, for either a personal search, property search or both. The required search before entering the premises may be just as important as the exit one. The use of alcohol and non-prescription drugs on company premises has reached alarming rates in many parts of the country. The entry search of the person and property will certainly reduce the flow of alcohol and drugs into the plant. Exit searches are made primarily to stop the unauthorized flow of parts and merchandise from the plant. Security guards usually have the responsibility of conducting the entrance and exit searches. It is recommended that all employees be notified that they will be subject to such searches and that the searches be done on a regular and equitable basis. All employees, office and plant, salaried and hourly, should be required to submit to the searches. Various electronic devices may be utilized to assist the security guards, especially if the parts or products are metal. If the parts or products are rubber or plastic, then different methods of detection must be employed. The use of closed-circuit television reduces the opportunity to steal, but the cost of buying, installing and staffing such a unit may not prove cost-effective.

The security procedures and devices utilized by any organization, whether it be a manufacturing plant, bank, retail store or hospital, must be directed to reducing the opportunity for crime. An organizational environment, regardless of its function, is not safe, nor is it secure, if it is subjected to employee theft. Yet, security procedures and controls which are too strict or harsh can have a negative effect on employees. Such programs should be balanced so that the negative effects on employee morale and productivity do not offset the recognized gains through reduced company loss.

3. *Improvement of Job Satisfaction.* During the last few years numerous studies have indicated that job satisfaction and morale are extremely important. Positive job satisfaction and morale have been linked to reduced levels of thefts and inventory shrinkage, with the opposite being the case with negative factors. The implication of works such as Abraham Maslow[8] and Frederick Herzberg[9] on the problem of thefts indicate that individuals operate on a "needs priority" basis, meaning that by eliminating an employee's financial "need" by paying a good wage should discourage theft. This might be an oversimplification of a major problem. However, the implications certainly suggest that any positive actions taken by the organization to improve job satisfaction will have a positive effect on the employee. Many recognized causes of employee thefts are reduced or eliminated when an organization takes positive action to improve job satisfaction. Even though job dissatisfaction may not directly relate to thefts, the root causes of both job dissatisfaction and thefts are the same or similar.

4. *Apprehension and Prosecution.* The policy of apprehension and prosecution is a viable control strategy, especially if all alternatives have been explored by the organization. This control strategy involves basically four aspects:

1) how must apprehension be handled in order to satisfy the legal requirements of an employee contract,
2) will the organization have a broad policy of prosecution in all cases,

3) will the company be willing or required to reinstate a person who has been detected and apprehended, but was acquitted in a court of law, and
4) will the organization consider restitution and continued employment as an acceptable means of deterring that employee and other employees in the future?

Any control strategy involving apprehension and prosecution must have as its foundation a solid security program. However, some organizations question whether the cost involved in a solid security program is worth the deterrent effect. The rate of prosecution of apprehended employee thieves is as low as 1% in some jurisdictions, and the decision to prosecute is not always up to the organization.

Still, the decision not to prosecute may often be for the best. A decision to prosecute requires a legally sound security program plus a willingness to withstand bad publicity, a willingness to defend against libel and malicious prosecution civil suits and false arrest cases, a willingness to bear the expense of having security employees, supervisors, managers and witnesses available in court for criminal prosecution of cases, and a willingness to support the employees involved in the cases, if a civil suit or complaint involves them in a personal manner. Several security authorities have stated that a decision to prosecute an employee caught stealing by one organization might not have a deterrent effect, whereas decisions to prosecute by all organizations certainly would have a strong deterrent effect.

An organization may be required, either by the statutes of a particular state or the employee contract that the organization has signed, to reinstate an employee who has been found not guilty to his original job with all back pay, raises and seniority which would have occurred during the absence. The fact that reinstatement may be possible, plus all of the expense and other ramifications, may cause an organization to want to have an apprehension and prosecution policy that will allow the organization to thoroughly review each case before deciding which course of action they will take.

A substrategy of the area of apprehension and prosecution is a company-imposed restitution program. Restitution may provide an opportunity to rehabilitate an experienced and productive employee. If an organization decides on this approach, two basic questions must be considered. First, who will be involved in deciding the penalties to be imposed? Second, will a restitution program be sufficient punishment and penalty to deter other employees from committing the same act? It is generally accepted that any restitution program must involve both labor and management in imposing penalties and overseeing the program. A restitution program which has been accepted by the employees certainly helps control present and potential violators. The question of the deterrent effect of a restitution program is more difficult. One must either look at the success or failure of other programs, or test the program for a period of time and evaluate the results.

Ed Stamper, Vice President of the National Business Crime Information Network, believes that any employee loss prevention awareness program must tap the potential and cooperation of every worker if such an undertaking is to be successful. People, according to Stamper, must be perceived as vital resources, absolutely essential to a successful loss prevention effort, Figure 11-2 illustrates the attributes of an effective employee loss prevention awareness/involvement program.

Two major problems which occur within most organizations and are indirectly linked to internal thefts are alcoholism and drug abuse.

Figure 11-2

Employee Loss Prevention Awareness Program[10]

An employee loss prevention awareness (involvement) program will:

1. Communicate top management/company-wide commitment to preventing abuses.
2. Communicate the primary objective of prevention, not detection.
3. Communicate exactly which activities are unacceptable.
4. Communicate *why* those activities are unacceptable, i.e., What are the mutual benefits of prevention:
   A. Business abuse harms employees, customers and suppliers.
   B. Business abuse hurts morale and productivity.
   C. Business abuse damages company's reputation.
   D. Business abuse creates financial losses which are passed on.
   E. Business abuse impacts everyone as consumers.
5. Communicate exactly what the concerned individual can do to help. This might vary from position to position. Starts with general guidelines and is expanded over time to focus on specific tasks.
6. Communicate the steps the company takes to prevent abuses.
   — Screening
   — Audits
   — Guards
   — Physical security
   — etc.
7. Solicits everyone's ideas, suggestions and information — always.
8. Starts at the *beginning of employment* and can be built upon throughout a career.
9. Includes a visible, easily accessible method of providing ideas, suggestions or information.
10. Periodically includes a "thank you" when no problems have occurred or improvements have been made. "The behavior you reinforce is the behavior you get."
11. Includes an appeal to customers, suppliers and neighboring businesses to help.
12. Be consistent company-wide — uniform to all levels of employees.

13. Address problems in an open manner, preventing the "grapevine" effect.
14. Be measured by long-term impact on company-wide attitudes and termed successful when a reduction in losses and/or problems occurs.
15. Be an ongoing part of the loss prevention effort. New themes, messages, and information are always necessary to maintain awareness. It is not a one-time project.
16. Have the understanding and support of on-site management.
17. Always address the individual as part of the solution, not a potential problem.

An employee loss prevention awareness (involvement) program will avoid:

1. Accusations or threats.
2. Focusing on internal theft or employee dishonesty. It should focus on the *acts* that cause loss not *who* commits them.
3. Being forgotten after the initial introduction.
4. The tendency to measure success by detection alone.
5. Being counted on to replace sound security measures.
6. Being counted on to replace communication from top management.

Source: Ed Stamper, *Loss Prevention Employee Awareness,* National Business Crime Information Network, 3688 Clearview Avenue, Suite 150, Atlanta, Georgia 30340.

Alcoholism

An alcoholic is defined as a person who cannot function on a daily basis without ingesting an alcoholic drink. Once a person reaches the point of excessive, compulsive drinking and dependence on alcohol, the person is considered an alcoholic and the condition a disease.

What are the problem indicators or progressive signs of alcoholism? The first indicator is generally that the person cannot stop after one drink, and so becomes drunk every time he takes a drink. Once a person has reached this stage, a significant change in personality is noticed after drinking; a person may get arrested for public intoxication, or domestic problems are likely to occur. This type of behavior is usually followed by drinking in the morning and sneaking drinks during the day. Frequent absenteeism is often one of the first signs that a supervisor notices about an alcoholic.

Why should an organization be concerned about the habits of alcoholics? First, the increased absenteeism that frequently accompanies alcoholism causes a decrease in production, and a decrease in production means lost earnings and profits. Second, an alcoholic cannot work at maximum efficiency if his/her body is suffering the effects of alcoholism. Third, an organization has a social and moral obligation (and sometimes a legal one, if it is so stated in the employee contract) to assist an employee with a problem.

Recognizing that alcohol was the most widely abused drug in America,

Congress passed the Alcoholism Act in 1970 and major amendments in 1974. This Act as amended established the National Institute on Alcoholism and Alcohol Abuse (NIAAA). NIAAA was given the responsibilty for formulating and recommending national policies and goals for the prevention, control and treatment of alcohol abuse and alcoholism, and for developing and conducting programs and activities to achieve the goals. This legislation also established an Occupational Program Area within NIAAA. This area has primary responsibility for alcohol abuse and alcoholism as they affect working people.

When a case of alcoholism has been discovered, the supervisor should take the following steps.

1) Present the facts to management and obtain approval to proceed with the case.
2) Obtain names, addresses, and telephone numbers for private and public counseling agencies (churches, Alcoholics Anonymous, etc.).
3) Discuss facts and counseling information with the employee and outline the exact procedures that the organization will follow. Explain exactly what is expected of the employee.
4) If the employee does not avail himself of the counseling service or continues to perform in the same unacceptable manner for an extended period of time, it is recommended that the supervisor process dismissal papers on the employee.

Step number four may seem harsh; however, an organization is not under any obligation to continue to employ an employee who refuses to alter his/her alcoholic habits. Hopefully, the action of the supervisor in step number three will be sufficient to cause the employee to face reality with the drinking problem.

Drug Abuse

Drug abuse is a serious problem for organizations as well as society. It is not a recent phenomenon, nor will it disappear in the foreseeable future. Opium and marijuana have been used for at least several thousand years, dating to the Egyptians in 2700 B.C. for marijuana and 1500 B.C. for opium.

Congress recognized the necessity of dealing with the problem of drug abuse and passed the Comprehensive Drug Prevention and Control Act in 1970. In 1973, drug enforcement efforts were organized into one single agency, the Drug Enforcement Administration, under the authority of the Department of Justice.

A drug is generally defined as a substance which, by its chemical composition and action, alters the structure or function of a living organism. Drugs are generally divided into four classifications: (1) narcotics, (2) stimulants, (3) hallucinogens and (4) depressants or sedatives.

Narcotics are opium and its derivatives, generally morphine, heroin, codeine or methadone, and are used legally to induce sleep and relieve pain. Heroin is the most common illegally used narcotic in the United States.

A stimulant is a drug which affects and acts on the central nervous system by increasing the activities of specific areas of the nervous system. A person affected by stimulants is usually excessively active, nervous, and talkative. The most widely used stimulant is caffeine, found in coffee, tea and cola drinks. Other commonly known and used stimulants are cocaine and amphetamines.

A hallucinogen is a drug which produces dream images and psychedelic reactions in an individual. Hallucinogens are generally recommended for clinical use and research and do not have any recognized medical utility. A person affected by hallucinogens may exhibit irrational behavior and may appear to be frightened or in a trance. The most widely used hallucinogen is marijuana, the active ingredient of which is tetrahydrocannabinol (THC). Other hallucinogens are lysergic acid diethylamide (LSD), methyl dimethoxy (STP), mescaline, and D-methyltryptomine (DMT).

A depressant or sedative is a drug which affects the central nervous system, slowing a person's actions and calming anxieties. Sedatives have a wide variety of medical uses, notably to control hypertensive activity. A person affected by depressants is usually disoriented, staggers and has difficulty concentrating. The most common and widely used depressants are barbiturates, of which phenobarbital and secobarbital (Seconal) are among the most recognized.

In any organization, the supervisor has primary responsibility for evaluating the conduct and performance of a subordinate on a daily basis. It is recommended that the supervisory procedures for dealing with an alcoholic be utilized by the supervisor in the case of a habitual drug user.

Summary

Internal thefts and crimes represent one of the major ways that an organization loses supplies, parts, equipment, merchandise, money, and organizational vitality. Statistics indicate that the employee represents a greater threat to the organization than the more publicized robber or burglar. Internal theft prevention programs cannot be viewed in a one-dimensional manner: they must involve an extensive, in-depth evaluation of the organizational environment and cultural contrasts; they must utilize the potential of the workforce to solve its own problems; and they must be conducted so that each activity interfaces with and relies upon the strengths of the other program activities.

Discussion Questions

1. What is embezzlement and how significant is it today?
2. What are the human and physical elements generally present in an embezzlement?
3. What are the various strategies for reducing and controlling internal theft losses?
4. Briefly discuss the uses and reliability of the Psychological Stress Evaluator (PSE).
5. What are some of the procedures which can be used to deter internal theft?
6. Discuss some characteristics of a business environment with high fraud potential.

Notes

1. Ed Stamper, *Loss Prevention Employee Awareness.* National Business Crime Information Network. Atlanta, GA 30340.
2. Ibid.
3. Richard Hefter, "The Crippling Crime," *Security World,* March 1986, p. 38.
4. Ibid., pp. 36-38.
5. "Embezzlement" in *Black's Law Dictionary.* 4th Ed. (Henry Black, ed) St. Paul, Minn., West Publishing Co., 1951, p. 614.
6. Black, p. 788.
7. Jack Bologna, *The Fine Art of Fraud Auditing.* Computer Protection Systems, Inc., Plymouth, Michigan, pp. 98-102.
8. Abraham Maslow, *Motivation and Personality.* (New York: Harper and Row), 1954.
9. Frederick Herzberg, *Work and the Nature of Man.* (New York: Thomas Y. Crowell, 1966.)
10. Ed Stamper, *Loss Prevention Employee Awareness.*

Chapter 12
External Threat and Crime

The previous chapter dealt with crimes which are usually committed by an employee. Even though the crimes committed by employees generally have a far greater impact on the organization than do crimes committed by individuals outside the organization, these external threats still represent enormous dollar losses.

Robbery

Robbery is the taking or attempting to take anything of value from the care, custody, or control of a person or persons by force, threat of force or violence, or by putting the victim in fear.

In 1985, the estimated robbery total in the United States was 497,874, or one every 63 seconds, with an average value loss per robbery of $628 for a total reported loss of $313 million. Nationally, nearly half of the reported robberies committed in 1985 were perpetrated on the streets or highways. Among the various categories, bank robberies comprised less than 2% of all robberies but registered the highest average loss: $3,048 per incident.

Most experts agree that robbers plan their crimes to some extent. The first step would be to make a decision to commit a robbery. The robber usually wants easy money quickly and will look for a target which offers little resistance and a minimum of exposure and risk. Any business located in a remote area or neighborhood, operating at times of the day or night with little or no pedestrian or street traffic, easy to enter and exit and having only one employee is a business that is very vulnerable to a robbery. For example, a robbery target with most of the above conditions would be a convenience store or gas or service station located on a major street or thoroughfare, staffed by a single employee, operating late at night with a considerable amount of cash in order to make change.

Apart from the loss of money, the most serious aspect of a robbery is the threat to the operator or employee's life. Statistics indicate that as many as one out of five commercial robbery victims suffer injury or death at the hands of robbers. In those cases that could be verified, the person being robbed offered some degree of resistance in a majority of the injury or death cases.

What can be done to reduce the risk to the operator or employee during a robbery? The most important thing that can be done to reduce the risk during a robbery is training in a few basic rules of safety. Operators and employees should:

1) Do whatever it takes to stay alive. Money or merchandise will not have any importance if you are dead.
2) Reassure the robber that you will cooperate in any way possible.
3) Remember that once a robbery is in progress, it is too late for preventive measures.
4) Remain as calm as possible.
5) Remember and concentrate on what the robber wants and what you are to do to satisfy the robber.
6) Discreetly observe the robber and look for general characteristics of height, weight, race, sex, weapon, clothing and mode of travel as well as specific characteristics like tatoos, scars, missing teeth, and speech impediments.
7) As soon as the robber has left the premises and you are safe, immediately call the police. Do not notify anyone else before calling the police.

These instructions apply only to a situation after a robbery has started. What should be done by a business to reduce the likelihood that a robbery will be committed at their location? First, the amount of cash should be kept as small as possible. Making daily or timely bank deposits and placing unneeded cash in a safe which cannot be opened by the employee are common and acceptable practices. Advertise this practice with signs both outside and inside the premises. Second, operate the business with at least two employees. There may be businesses which cannot economically afford to operate at all times with two employees, but even if this is not possible, the business should always be opened and closed with two employees. Third, money should not be taken to and from the business to the home of the operator or employee. This policy gives a robber two targets. Fourth, the operator or employee should not open the business until the operating money has been taken from the safe and all procedures for doing business are ready. Preparations for doing business should always be limited to non-money types of actions, such as stocking a counter. Fifth, cash registers should be located in front of the building near the entrance and exit doors, and in view of pedestrians, passing motorists and police officers. Windows and doors should be unobstructed so that it is possible to view the check-out area and cashier. Sixth, have a telephone available on the premises at a location other than the cashier area. One of the techniques which robbers use is to disable the telephone located near the cash register. Seventh, the lighting should make it possible to see both inside and outside the premises. At night, it is useful for the operator or employee to observe and recognize customers as they enter the business. This will aid in preventing both robberies and shoplifting.

In order to reduce the crime of robbery, a business must plan and implement robbery prevention policies and procedures. Robbery prevention not only saves dollar losses but may save lives or prevent serious injuries.

Burglary

Burglary is the unlawful entry of a structure to commit a felony or theft. Burglary is usually categorized as: forcible entry, unlawful entry where no force is used, and attempted forcible entry. Burglary is the second most prevalent property crime in the United States. In 1985, 3.1 million burglaries were reported, or one every 10 seconds. Seventy percent were forcible entries, 22 percent were unlawful entries without any force, and 8 percent were forcible entry attempts.

Burglary represents a substantial financial loss to the victims. In 1985, burglary victims suffered losses totaling $2.9 billion, and the average dollar loss per burglary was $953.

Both the residential and nonresidential burglary categories showed 3 percent increases, 1984 versus 1985. Daytime residential and nonresidential burglaries were up 4 percent. Burglaries committed during the nighttime hours increased approximately 1 percent for both nonresidential and residential property.

As with robbery, the impulse to commit a burglary begins with the need or desire for money or merchandise. Reducing the likelihood that a burglar will select a particular target is a complex problem. For most businesses, the best overall strategy is to ensure that likely entry points are safely guarded by physical and visible means of discouragement like lights, locks, fencing, and alarms, making entry appear noticeably difficult. This is called "target hardening." The second step is to be sure that the lighting, fencing, locks and alarms will make it difficult and time consuming to burglarize the building. Any delay in entering a building in a lighted area will increase the risk of being detected and apprehended. The third step is the use of a silent alarm, activated by intrusion, which signals to some outside respondent (usually an alarm company) that a break-in is in progress. The fourth preventive step involves marking merchandise in a way that is impossible or difficult to remove. This makes the merchandise harder to fence, and easier to trace and identify.

Burglars are 99 percent male and can be placed into three categories: the rank amateur, the semi-professional, and the professional. Rank amateurs comprise approximately 70 percent of all known burglars and usually are opportunists looking for the poorly defended target. Any of the above-mentioned preventive steps will make a target too difficult for the amateur. Semi-professional burglars comprise more than 28 percent of known burglars and pose the greatest immediate threat to a business. This person will not be easily deterred by a weak combination of lights, locks and alarms, and generally has the means to dispose of stolen merchandise. The professional burglar comprises less than 2 percent of the known population of burglars and poses the greatest long-term threat to a business. This person can pick locks, bypass alarms, turn off lighting systems and open safes. The professional burglar is usually only interested in obtaining extremely large amounts of money or valuable merchandise which is small in size. Thus, he can be deterred by reducing the amount of money or merchandise available on the premises.

In the event that a burglary occurs and immediate apprehension is not possible, the police department should be called before entering the premises. The police use the "transfer theory," which states that a person brings something to the crime scene with him and takes something when he leaves. If the police are able to identify evidence associated with either part of the transfer theory, it greatly increases their chance of success in solving the crime.

Shoplifting

While no type of retail business is entirely immune from shoplifting, the retail stores most frequently affected by shoplifting are those selling small and concealable items. Grocery, hardware, drug and general merchandising stores are the most vulnerable in terms of items that are easy to shoplift.

For a number of reasons, estimates on the actual extent of shoplifting losses are difficult to determine. Questions regarding who, when, where, why and how retail property was stolen usually go unanswered. Generally, there is no factual determination of whether an employee or a shoplifter took the merchandise, whether it was lost in shipment, or if it was a paperwork error of recording or processing. Various studies and reports have indicated enormous variability regarding the range and scope of shoplifting. Perhaps the most precise shoplifting data is that provided by the Federal Bureau of Investigation, Uniform Crime Reports. However, this annual report is indicative of only reported crimes and is not reflective of number of crimes actually committed or total losses.

The UCR for 1985 reported a total of 970,000 shoplifting incidents handled by the nation's police departments. Total losses were reported to be $82.5 million with a per incident average of $85. However, a single study of shoplifting, conducted by Commercial Services Systems, Inc., of Van Nuys, California, a loss prevention and market research firm, estimates that the U.S. grocery industry alone loses more than $2 billion a year to shoplifters. Thus, regardless of the data source, it is clear that retailers across America lose a tremendous amount of goods and merchandise to an army of shoplifters drawn from every segment of society. Bob Curtis, in his text, *Retail Security*, indicated that the ratio of shoplifting incidents to shoplifters apprehended may well range within the area of 100 to 1. That is, for every shoplifter caught, 99 others walk undetected out of the business with their stolen goods. How much is actually lost each year to shoplifters is open to conjecture, and better yet, how much more is the honest consumer paying for merchandise that must be priced higher to cover such losses?

In contrast to burglary, shoplifting occurs as a part of ordinary customer behavior, up to the point where the customer conceals merchandise or alters the price tag. In general, it is difficult to identify a particular item or define an approach that will characterize ordinary customers. Most shoplifters are amateurs who operate alone and are likely to steal at any time of the day or day of the week. Studies have shown that shoplifters like to operate when stores are most crowded or in a store where they believe that the employees are not trained and familiar with shoplifting techniques. Friday and Saturday are usually the big days for shoplifting, and the holiday seasons represent a time when stores are crowded and store personnel are very busy.

The behavior of shoplifters begins to differ from that of ordinary customers after merchandise has been removed from displays. First, shoplifters tend to concentrate on areas of the store where they cannot readily be observed from the cashier's area or by store personnel on the floor. Notably, they look for anything that obstructs vision or reduces visibility: high sales counters, tall displays, free-standing signs and poorly lighted areas.

A shoplifting technique used by shoplifting couples (called "rounders") is to take advantage of the above conditions to hide their shoplifting actions. These shoplifters operate in pairs in order to be able to assist their partner by blocking the view of store personnel and customers and concealing the merchandise.

Shoplifters are legendary for their ingenuity in hiding items. While most use purses, pockets or underclothing, some develop elaborate special-purpose equipment to increase their capacity to conceal merchandise. This equipment ranges from coats with extra pockets sewn into the lining to devices such as hooker belts that permit an array of merchandise to be hung around the shoplifter's waist under a bulky coat. Some even equip themselves with slings that permit them to carry comparatively large items such as radios or turkeys between their legs.

The two most serious problems associated with shoplifting are detecting the crime and dealing with the apprehended shoplifter. Most owners or managers agree that it is much better to have a program to reduce the opportunity for shoplifting than to place emphasis on the detection and apprehension phase of a complete shoplifting program; however, detection and apprehension of shoplifters must be a part of a comprehensive anti-shoplifting program.

There are three critical points which offer the greatest opportunity for control: (1) when potential perpetrators enter the store, (2) when they pick up the merchandise, and (3) when they leave the store. At the entrance, steps should be taken to intimidate shoplifters (e.g., signs warning of prosecution) or to reduce their capability to conceal merchandise (e.g., a system requiring customers to leave parcels with the cashier until they leave). Thereafter, when shoplifters are in the vicinity of their target merchandise, the courteous intervention of store employees offering assistance can put shoplifters on notice that they are being watched. Finally, when shoplifters have concealed merchandise and are preparing to leave, a mechanism (e.g., electronic sales tag sensor) that can either detect the presence of hidden goods or prompt the shoplifter to decide to pay rather than risk apprehension will be effective.

Bad Checks and Credit Card Fraud

Each year, businesses in the United States lose millions of dollars through bad checks. Most states classify bad checks as theft by deception, which covers checks written against no accounts, insufficient funds, forgeries or stolen checks. The dollar amount of the check determines whether the offense is a misdemeanor or a felony. Most criminal cases result from checks written on "no account," forgeries, or stolen checks.

Any business or organization which receives checks on a regular basis can do certain things to reduce the risk of cashing a bad check and of not being able to identify and locate the person responsible.

1) *Always insist on proper identification.* The best identification is one with a photograph, such as a driver's license. If a person does not have a driver's license, a secondary identification which gives certain specific information will suffice, i.e., a company identification card for an employee.
2) *Check cashing cards for a specific store.* Many stores require a person who wishes to cash a check to fill out an application and receive a check cashing card. Most stores which have this system will not cash a check without such a card.
3) *Compare name, address, and telephone number on check with identification.* Make sure to compare first, middle, last name, and address on check with the same information on the identification.
4) *Record identifying information from identification on check.* Record the driver's license number or other identifying number on

the back of the check.

5) *Check numerical amount against written amount.* A bank is obligated to honor the written amount on a check. Make sure that the two amounts are the same.
6) *Be cautious of "stale" dates; never accept post-dated checks.* Question a person about a check with a date that is more than one month old. Never accept a post-dated check.
7) *If signature is questionable, have the person sign again.* If the signature is illegible, and the person signs again and it is still illegible, have the person print his name.
8) *Verify account at bank.* The magnetic numbers on the lower left bottom of the check represents the banking information. The last set of digits are the account number. Numbers always begin exactly 5 and 5/8 inches from right corner.

Each business should establish a firm check cashing policy and advertise it. Strict dollar limits, amount of purchase only, and courtesy cards for check cashing are all good business practices. Post a sign to spell out your policy, and stick to it. Figure 12-1 is a form from the Kentucky Department of Justice, Office of Crime Prevention which may provide some assistance in receiving restitution for a bad check from the person who cashed it.

A check returned marked "no account" or "closed account" is a warning of extreme carelessness or fraud. Such a check is usually evidence of a swindle or fraudulent action, though in rare instances a customer may issue a check on the wrong bank or a discontinued account. If restitution is not received, prosecution would be warranted provided the check writer's identity can be established.

Forged checks are worthless. Any alteration, illegal signature, forgery of the endorsement, erasure or obliteration on a genuine check is a crime. Smudged checks, misspelled words, poor spacing of letters or numbers are an indication that such changes may have been made. A forged check issued to pay for merchandise or services received is a felony.

Credit card fraud is an area that is causing enormous dollar losses for certain business enterprises. For example, many banks are faced with greater dollar losses due to credit card fraud than they are from bad checks. At a recent security conference, a director of security for a large metropolitan bank stated that his bank was reassessing their policy on the issuance of credit cards, and that a major effort was being made to make sure that retailers checked on the card before accepting it. He also stated that a study was being conducted to see what kind of relationship exists between those skipped monthly payments and the eventual total loss.

It is recommended that any business that issues credit cards make a thorough credit check on the indivudal before issuing a credit card. At

Figure 12-1
Notice of Returned Check Form.

NOTICE OF RETURNED CHECK

Dear Customer:

You will find outlined below information relating to your check recently returned to us unhonored. We realize the complexity of keeping accurate bank balances and know that you would like to honor your check within ten (10) business days as allowed by Kentucky law. We appreciate your business and are sorry for the inconvenience.

10 Day NOTICE **RETURNED CHECK INFORMATION** **10 Day NOTICE**

You are hereby notified that your check dated __________ 19__, for $_______ has been presented to the bank for payment and has been returned to us unpaid. Please arrange to pay the amount of this check within ten (10) business days from the date you first receive this notice.

Date ____________________ ____________________ (Store)

Signed ____________________ ____________________ (Address)

Phone ____________________ ____________________ (Town, State)

(PLACE CHECK HERE WHEN COPYING)

Sections 514.040 and 514.090 of Kentucky Penal Code state in part that ". . . an issuer of a check or similar sight order for the payment of money is presumed to know that the check or order, other than a postdated check or order, would not be paid, if:

(a) The issuer had no account with the drawee at the time the check or order was issued; or

(b) Payment was refused by the drawee for lack of funds, upon presentation within 30 days after issue, and the issuer failed to make good within 10 days after receiving notice of that refusal."

Conviction of an individual under the provision of the Penal Code concerning bad checks is a Class A misdemeanor unless the value of the property is $100 or more, in which case it is a Class D felony. A Class A misdemeanor is punishable by a fine up to $500 and a possible term of imprisonment not to exceed twelve (12) months. A Class D felony is punishable by a possible term of imprisonment of not less than one nor more than five years and a fine not to exceed $10,000 or double the offender's gain from commission of the offense, whichever is greater.

IMPORTANT—Please make your check good within 10 days! You will notice that Kentucky law states that an issuer of a check is presumed to know that the check would not be honored by the bank when the issuer fails to make the check good "... within ten (10) days after receiving notice of that refusal."

If you fail to honor your check within 10 days after receiving this notice we will be forced to consider appropriate legal action.

least one card company requires retailers to check the card before the transaction, similar to a bank verification on a check. These two processes will greatly reduce the dollar loss to credit car fraud.

Current estimates seem to indicate that losses due to credit card fraud are approximately one billion dollars each year. Credit cards are often fraudulently obtained from issuers with the intent to defraud. They are often stolen from mailboxes, homes, offices, and from individuals, and then used to obtain merchandise from an unknowing merchant. Other methods of fraudulent use include using a counterfeit or altered credit card, using a credit card previously canceled by the issuer, and providing a false credit card number. When the credit card of another is illegally obtained and used to effect transactions, the crime of forgery is committed.

Methods of preventing credit card fraud in retail operations are similar to the techniques for protection against bad checks. Store personnel should use "hot lists" (issuers' list of lost or stolen cards), be aware of established purchase limits, and insist upon proper user identification.

Bombs and Bomb Threats

The target of a bomb is not usually selected at random. Target selection is usually based on some type of gain or revenge for the person or persons responsible.

The explosive devices are generally set for detonation at a time sufficient for the person responsible to be a considerable distance away before the bomb-threat call is made and the explosive device detonated. It is, however, becoming more common for terrorist groups to detonate a device at a specific location and announce that a particular group is responsible at a later date and/or time.

Management, especially of large multinational corporations, must consider bombs and bomb threats as a serious problem. Three factors must be examined in any bomb or bomb threat policy: the degree of risk and danger to personnel of injury or death, the consequences of damaging or destroying materials in a facility, and the total cost or loss from a bombing or a bomb threat. The risk factors associated with bomb threats must be balanced against the consequences of evacuating a facility every time a threat is received. A preplanned cursory search may be a more acceptable alternative than evacuation in a situation where the known risks are not great.

Bomb Threat Guides

If a business receives a bomb threat, the person receiving the call should try to obtain as much information about the bomb threat as the caller will give. (Figure 12-2)

In most cases, however, the caller will not remain on the telephone for a long period of time or answer questions. Second, the person receiving the call should notify the Director of Security or the person responsible for

Figure 12-2
Bomb Threat Guide.

Time and Date Reported: ______________________

How Reported: ______________________

Exact Words of Caller: ______________________

____

____

Questions to Ask: ______________________

1. When is bomb going to explode? ______________________
2. Where is bomb right now? ______________________
3. What kind of bomb is it? ______________________
4. What does it look like? ______________________
5. Why did you place the bomb? ______________________
6. Where are you calling from? ______________________

Description of Caller's Voice: ______________________

Male _____ Female _____ Young _____ Middle Age _____ Old _____ Accent _____

Tone of Voice ________ Background Noise ________ Is voice familiar? ________

If so, who did it sound like? ______________________

____

Other voice characteristics: ______________________

Time Caller Hung Up: ________ Remarks: ______________________

____

____

____

Name, Address, Telephone of Recipient: ______________________

____

____

security at the facility. The Director of Security will make a decision to call the police, fire department, or both and to search and possibly evacuate the facility. A decision to search will generally be done according to a prearranged plan and may be limited to specific areas of the facility, depending on the information given by the caller. (Figure 12-3)

A decision to totally evacuate a facility is usually made only when there has been a history of having found bombs either in the area or at the facility, or when the caller gives enough information to the receiver to warrant an evacuation. The tone or urgency of the conversation and the possible identity of the caller, i.e., a teenage caller with giggling friends

Figure 12-3
Bomb Search Systems.

	SEARCH SYSTEMS			
	SEARCH BY: Supervisors	ADVANTAGES	DISADVANTAGES	THOROUGHNESS
SUPERVISORY	BEST for Covert search POOR for thoroughness POOR for morale if detected	1. Covert 2. Fairly rapid 3. Loss of working time of supervisor only	1. Unfamiliarity with many areas 2. Will not look in dirty places 3. Covert search is difficult to maintain 4. Generally results in search of obvious areas, *not* hard-to-reach ones 5. Violation of privacy problems 6. Danger to unevacuated workers	50-65%
OCCUPANT	SEARCH BY: Occupants BEST for speed of search GOOD for thoroughness GOOD for morale (with confidence in training given beforehand)	1. Rapid 2. No privacy violation problem 3. Loss of work time for shorter period of time than for evacuation 4. Personal concern for own safety leads to good search 5. Personnel conducting search are familiar with area	1. Requires training of entire work force 2. Requires several practical training exercises 3. Danger to unevacuated workers	80-90%
TEAM	SEARCH BY: Trained Team BEST for safety BEST for thoroughness BEST for morale POOR for lost work time	1. Thorough 2. No danger to workers who have been evacuated 3. Workers feel company cares for their safety	1. Loss of production time 2. Very slow operation 3. Requires comprehensive training and practice 4. Privacy and violation problems	90-100%

Source: U.S. Government Printing Office: 1981 0 - 339-713.

in the background, will provide a clue as to the credence of the call. In most cases, a decision to evacuate on a limited basis is both practical and feasible.

Terrorism

There are varying definitions of terrorism, which is due in part to the fact that acts of terrorism are perceived differently by the victim, the perpetrator, and by others who are not directly involved as victims or perpetrators. For example, disorders and terrorism have both common characteristics and specific differences. Both are forms of extraordinary violence that disrupt the civil peace; both originate in some form of social excitement, discontent, and unrest; both can endanger massive fear in the community. Disorders and terrorism constitute, in varying forms and degree, violent attacks upon the established order of society. However, the focus, direction, application, and purpose of the terror and fear are different.

Disorders do not necessarily have political overtones, as they may arise simply from excessive stimulation and exuberance during an event such as a football game or rock concert. Disorders are often haphazard events with no planned happenings or directed objectives. They are collective discharges in human behavior that may range from abusive language and resistance to destruction of life and property.

Terrorism is characterized by planned, calculated acts aimed at manipulating society toward defined objectives. The message of fear is deliberate and is the very purpose of the terrorist activity. Without an audience, terrorism is an exercise in futility.

The difference between terrorism and "terrifying" criminal acts lies in the technique and purpose of the violent act. Terrorism is coercive, designed to manipulate the will of its victims and its larger audience. Similarly, the fear generated by a criminal act such as rape is aimed at overcoming the will of the instant victim, not at the minds or resistance of others. Thus, a criminal act must be qualified before it can be coined "terrorism;" however, any act of terrorism would in every case be considered a criminal act by someone.

The Private Security Advisory Council's Committee on the Prevention of Terroristic Acts defined terrorism as "criminal acts and/or threats by individuals or groups designed to achieve political or economic objectives by fear, intimidation, coercion, or violence." Further, the committee identified five distinctive categories or kinds of terrorism: (1) political terrorism; (2) nonpolitical terrorism; (3) quasi-terrorism; (4) limited political terrorism; and (5) official or state terrorism.[1]

"Political terrorism" in its fully developed form is revolutionary in character; whether it is a realistic tactic or not, it has as its purpose the subversion or overthrow of an existing government. It is characterized by: (1) its violent, criminal nature; (2) its impersonal frame of reference; and (3) the primacy of its ulterior objective, which is the dissemination of fear throughout the community for political ends or purposes.

"Nonpolitical terrorism" includes a vast area of true terroristic activity that clearly cannot be termed political. Such terror may affect society and its patterns of behavior on a considerable scale, but the objectives of those involved are individual or collective gain, rather than the achievement of political ends. Organized crime, street gangs, and "Charles Manson" type groups characterize some forms of nonpolitical terrorism.

"Quasi-terrorism" is the use of terroristic techniques or tactics in situations that are not true terroristic crimes, such as the taking of hostages during the course of a conventional crime such as bank robbery. Another example is skyjacking, where the aircraft and hostages are threatened subject to the payment of ransom for private gain.

"Limited political terrorism" involves acts of terrorism which are committed for ideological or political motives but are not part of a concerted movement to capture the government. Singular acts committed by a lone terrorist or acts limited to a particular social context constitute limited political terrorism.

"Official or state terrorism" involves situations where opposition between the oppressive policies of a state (government) and the will of the people leads the state to use violence and this reciprocally causes the people to react by violent means. Frequently, terroristic behavior by individuals and dissident groups is claimed to be a response to acts of terrorism sponsored by the state.

Thus, terrorism is a serious and complex problem of national and international concern that defies easy, conclusive solutions. Acts of terrorism, including assassinations, kidnappings, bombings, and personal assaults, are frequent and widespread. Modern terrorists have been assisted by the ease of intercontinental travel and mass communications. Acts of terrorism have gained immediacy and diffusion through television, which conveys the terrorist message to millions worldwide. The terrorist of the 80's has been magnified, enlarged beyond his own powers by others.

According to Risks International, a terrorist watch group, forty terrorist attacks were made against U.S. businesses in 1985, killing two Americans, injuring two, and causing $1.72 million in damages. Additionally, over $6 million in kidnap ransom was demanded of U.S. businesses.

Of the total 3,010 terrorist incidents in 1985, U.S. interests were targeted approximately 100 times. The types of attack included 1,527 bombings, 990 attacks against facilities, 374 assassinations, 109 kidnappings, and 10 hijackings. The results: 7,166 deaths, 5,181 injuries, $46 million in damages, and $17.6 million in ransom demands. Businesses, including utilities and transportation, were targeted 986 times; police and military personnel or installations, 946; government facilities, 529; and political parties, 117. Latin America was the foremost terrorist area with 1,933 incidents, Europe had 471, Asia 326, the Middle East and North Africa 171, the Sub-Sahara 99, and North America 10.[2]

It is obvious from the 1985 data provided by Risks International that North America is not yet a hotbed of international terrorism. However, the experiences of the U.S. government and international firms in many countries of the world leave little doubt that American interests and people are viewed as primary targets. The potential for increased terrorist activity within the United States is significant, and, according to many experts, will be on the rise in the years ahead.

Executive (Personnel) Protection

The threat of a kidnapping or an extortion attempt varies from area to area, both in the United States and other countries of the world; therefore, threat assessments of the likelihood of their occurrence must be definitive and specific to localized situations. The political climate and regional conditions around company facilities and areas frequented by company personnel should be determined and documented. Since the risk level and exposure of company executives and employees will vary, the most vulnerable facilities and personnel should necessarily be afforded a higher degree of security.

Given the occurrence and frequency of political kidnappings and hostage taking, it is obvious that extremists view such acts as an effective tactic toward accomplishing specific goals. Political extremists and/or terrorists utilize the violence and suspense of the kidnapping or hostage situation to cause the public to recognize their power and exert pressure on those in authority to comply with their demands. Such incidents attract the attention of the public and the media, and require minimal commitments of terrorist personnel and equipment.

Thus, affirmative actions to prevent or discourage kidnappings or extortion attempts must be undertaken prior to the occurrence of such incidents. An organized effort of planning and preparation must consider the fact that any response to terroristic acts requires policies, resources, analysis, communications, and decision making that go beyond the usual boundaries of a single organizational function. The concept of the Crisis Management Team offers a means to draw together the necessary areas of management to obtain a viable, defensive strategy.

Crisis Management

In 1976, ten multinational corporations began funding the development of a "Crisis Management System," and Motorola Teleprograms, Inc., later made the results of the study available to other organizations. The results of the study concluded that a Crisis Management Team (CMT) should be formulated by any business facing the threat of terrorism. The CMT would have eight basic functions in regard to preparing for, and reacting to, terrorism:

1) **Leadership,** which focuses the input and direction from senior management through a CMT person.
2) **Security,** which supplies expertise and experience with a systems approach to assets protection and a knowledge of adversaries to the business.
3) **Legal,** which provides information on the legal implications of CMT strategies and tactics.
4) **Financial,** which develops the monetary base for CMT, record and document expenditures.
5) **Personnel/Medical,** which provides the necessary health and biographical data on personnel and their families in the corporate framework.
6) **Public Relations,** which will provide the expertise to assess the full impact of a crisis and respond to the media.
7) **Adversary Communications,** which will exchange information with the terrorists.
8) **Crisis Counsel,** which advises on terrorist groups and the techniques they use to victimize organizations. Crisis counsel will act as the "devil's advocate" for the organization's terrorism plan.[3]

Thus, crisis management is a dynamic concept that dramatically increases a company's efficiency and effectiveness in handling the many incidents that affect organizations. It applies sound managerial techniques designed to cope with a crisis situation while simultaneously accomplishing organizational goals.[4]

Preparing for the eventuality and occurrence of terrorism requires much planning and foresight. It is most important that potential victims be cognizant of the depth of skill and commitment that terrorists have so aptly demonstrated to their worldwide audience.

Espionage and Sabotage

Some corporations worth many millions of dollars owe their existence to trade secrets than can be written on a few sheets of paper. Certainly every corporation has some trade secrets or confidential information that it wishes to keep out of the hands of competitors. This is especially difficult for international corporations, because trade secrets illegally obtained in this country may be perfectly legal in a foreign country which does not recognize our laws pertaining to such incidents. An American-controlled multinational corporation assembled through a series of mergers may find that it is practically impossible to control the actions of a foreign component. Any trade secrets held by a foreign subsidiary may be compromised by disloyal employees. Multinational corporations have employees from a variety of nations. The loyalty of such an employee would be suspect when compared with loyalty for his/her native country. Finally, most equipment used in business espionage, such as cameras, copy machines, and electronic listening and recording devices, are not licensed or registered. The buying, possession and use of these types of equipment would not place a person in a suspect position even in many foreign countries. Thus, the control and security of organizational information is of vital importance, and steps must be taken to ensure that its confidentiality is uncompromised. Its classification and subsequent protection as described in Chapter Eight are necessary to negate potential occurrences of espionage.

A business must be concerned with two types of sabotage: (1) actions which could cause disruption of vital services, and (2) actions which threaten or may threaten human life. Many acts of sabotage such as disruption of communication or computer systems may not directly affect human life; however, other actions such as arson or bombing may directly affect human life.

Sabotage can be defined as the willful destruction of property. The saboteur, whether a trained professional or an amateur, may be anyone within or without the organization. A saboteur may work alone or in concert with others. There are numerous methods and devices used to accomplish sabotage. Generally, the techniques of sabotage can be classified as follows:

Explosive Devices—Various explosive devices (bombs) containing either low or high explosives can be used to partially or totally destroy a facility or a vital component of it.

Mechanical Sabotage—Motors and machinery can be rendered inoperative by a variety of methods: breakage, insertion of abrasives into machinery, adding contaminants to fuel, acts of omission (failure to do proper maintenance), and placing false information on maintenance charts.

Fire—Fire is one of the oldest and simplest forms of sabotage. Whether the saboteur sets a simple fire or uses a timing device, fire is one of the most effective methods of sabotage because it can destroy both the property and evidence of criminal action.

Electrical (Public Services)—Almost every facility is dependent on public services for its continued operation. Interruption of these can be accomplished in a variety of ways. For example, electrical power lines, substations or generating facilities can be rendered inoperative by cutting lines, bombing generating or substation units, or by mechanical means.

Chemical—Chemical sabotage can be accomplished by the introduction of various materials into fuel supplies, service water, or drinking water. The magnitude and effect of this method of sabotage is limited only by the sophistication of the saboteur.

Psychological—Creating a negative situation or confrontation between employer and employees can often accomplish the objectives of the saboteur. While property is not destroyed, the employer's ability to produce may be disrupted or destroyed. For example, work slowdowns, walkouts, or general strikes can be instigated by one individual bent on creating unrest and disruption.

Sabotage is, thus, a real threat to government, industry, and even to individuals who have been targeted by someone wishing to cause them harm. Whether the motivation is criminal, political, or personal (revenge, mental imbalance, etc.), the results are the same.

Summary

External threats and crimes, although not representing as large a dollar loss as internal threats and crimes, are very important to businesses. External threats and crimes also involve a greater risk of injury, potential injury or death than do internal threats and crimes.

Many businesses today operate in areas where research and technology are changing their operations at a rapid pace. A competitive edge in one of the high technology areas may be easily lost through crime. In fact, the potential for a financial disaster from a single incident of espionage or sabotage may be greater than from a crime of any other kind.

Discussion Questions

1. Outline what a victim should do during a robbery.
2. What are some of the preventive measures which reduce the occurrence of burglaries?
3. What is the transfer theory and in what cases is it used?
4. What are some of the steps which an employee should take when accepting checks?
5. What are the steps that a person should take when receiving bomb threats?
6. Outline the types of terrorism.
7. What is crisis management?
8. What are some of the essential elements of a bomb threat guide?

Notes

1. *Task Force Report on Disorders and Terrorism*, U.S. Government Printing Office. Washington, D.C. 1976, pp. 3-6.
2. Risks International, Inc., Arlington, Virginia.
3. Lloyd W. Singer and Jan Reber, "A Crisis Management System," *Security Management*. September, 1977, p. 7.
4. Robert F. LittleJohn, "When the Crisis is Terrorism," *Security Management*. ASIS, August, 1986, p. 38.

Chapter 13
Fire Prevention and Safety

The management of an organization has a responsibility to develop and implement programs in security, safety, and fire prevention. While the duties and responsibilities of these three areas are different, the goal is the same: to prevent, reduce, or eliminate company loss. The functions of these areas are often combined into one major department, or they may operate as individual, cooperative units within the company. Major corporations frequently staff separate security, fire, and safety departments, all under a corporate director of loss prevention. However, regardless of the administrative arrangement, the role of security within an organization requires that security personnel be familiar with the techniques of fire prevention and control and employee safety.

Characteristics of Fire

Fire is one of the major threats to life and property in any business, and all security employees should therefore be aware of the fundamentals of fire and fire protection.

The Triangle Concept of Fire

Fire is defined as rapid oxidation, accompanied by heat and light. Oxidation is the chemical union of a substance with oxygen. The rusting of iron is oxidation, but it is not fire because no light and only a little heat is generated.

Fire can only occur when three things are present in sufficient quantity: oxygen, fuel to combine with the oxygen, and heat sufficient to maintain combustion. The removal of any one of these three factors will result in extinguishment of the fire. The classic fire triangle illustrates this by representing each of the three factors as the three sides. (figure 13-1)

The fire triangle is a simplified illustration of the elements necessary to initiate and sustain a fire. A more scientific and thorough explanation is a concept called the tetrahedron of fire. This explanation of fire adds a fourth element designated as "uninhibited chain reactions." This is an expansion of the "heat" side of the fire triangle, because there are a limited number of sources of activation energy other than heat which can, given the right conditions, initiate the chain reaction necessary to cause a fire.[1] The following discussion illustrates both the "fire triangle" and "tetrahedron" concepts of fire.

Figure 13-1
The Fire Triangle.

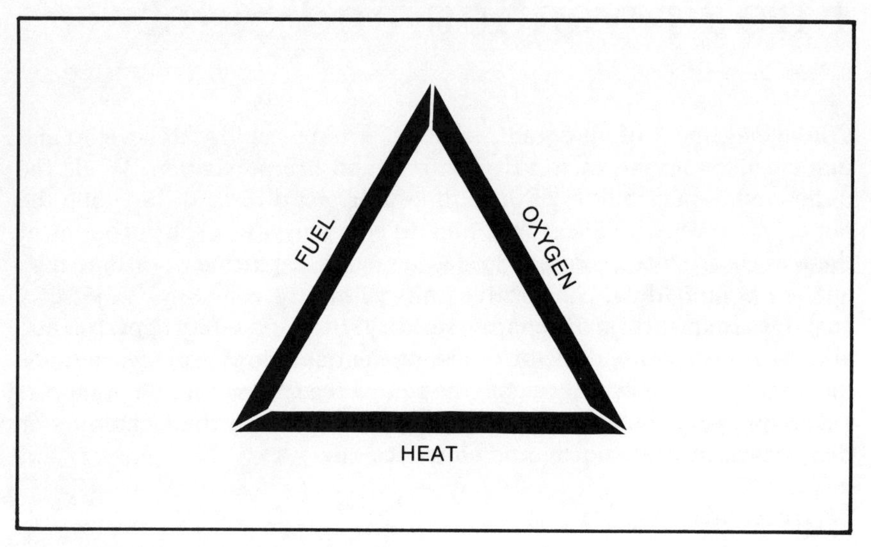

Oxygen

Most fires draw the needed oxygen from the air. If a fire burns in an enclosed area, the oxygen is gradually used up and the fire will diminish, often to the point of going out. If a limited but continuous supply of oxygen is present the fire may enter a smoldering stage. Since oxygen is readily available from the air, eliminating it as a factor of combustion is most often impractical. But in efforts to suppress an ongoing fire, various extinguishing agents can be used to "smother" the fire.

Heat

Heat for ignition can come from many sources: the sun, lightning, open flame, friction, electrical sparks and so on. In most circumstances a fire occurs when a source of sufficient external heat comes in contact with a combustible material, or in some cases, the material itself creates enough internal heat to trigger combustion (spontaneous ignition). The degree of heat necessary to start combustion varies depending on the type of fuel present. For combustion to occur, most materials must be heated rather rapidly, and once ignition temperature has been reached burning will continue as long as the fuel remains above this temperature. Heat of combustion varies with every type of fuel and is usually expressed in BTUs

(British Thermal Units). A BTU is the amount of heat required to raise the temperature of one pound of water one degree Fahrenheit at sea level.

The heat side of the fire triangle offers many ways to prevent and extinguish fires. One way is to keep all sources of ignition away from material to be protected, but if heat is required nearby, one may apply safeguards to the heating devices or insulate the materials. Extinguishment of a fire involves lowering the temperature of the fire to a point below the fuel's ignition temperature. The most common extinguishing agent used to accomplish this cooling effect is water.

Fuel

Fuels may be in a gaseous, liquid or solid form, but combustion normally occurs when a fuel is in the gaseous or vapor state. Solids and liquids, therefore, must vaporize before oxygen can react with the fuel in combustion. There are exceptions, but they are unique and limited. Most ordinary combustible solids are compounds of carbon, hydrogen, nitrogen, and oxygen along with other smaller portions of other minerals. When free burning in air occurs, oxygen reacts with carbon to form carbon dioxide and with hydrogen to form water vapor. The minerals and nitrogen compounds usually remain in the solid state as ash.

When a material in liquid form is heated, it will reach a temperature above which it cannot go and still remain a liquid. This temperature is called the boiling point. It is at this stage that liquids vaporize and mix with the air. The chance of a fire starting in a given flammable liquid and the speed of its combustion is dependent upon its particular characteristics and environmental conditions.

Fuels in the gaseous state are broadly divided into two groups: those that are flammable and those that are not. Flammable gases are perhaps the most difficult fuels to deal with because many are colorless and odorless. For safety and fire prevention, all gases should be stored in airtight containers or pipes.

To summarize, the triangular concept of fire and fire prevention rests upon the following principles:

1) An oxidizing agent (oxygen), a combustible material (fuel), and an ignition source (heat) are essential for combustion.
2) The combustible material must be heated to its ignition temperature before it will burn.
3) Combustion will continue until:
   (a) combustible material is removed or consumed,
   (b) oxidizing agent concentration is lowered to below the concentration needed to support combustion, or
   (c) combustible material is cooled to below its ignition temperature.

As a fire condition develops and starts, four distinct stages of the combustion process are involved, progressing from one stage or level to the next.

1) *Incipient stage*—At this initial stage, invisible products of combustion are given off; the oxidation process has begun. There is very little heat at this stage and neither flame nor smoke is visible.
2) *Smoldering stage*—Smoke is visible at this stage, but there is still little appreciable heat or flame.
3) *Flame stage*—Flame and smoke are both visible, but excessive heat is still not apparent.
4) *Heat stage*—Almost simultaneous with the flame stage is the generation of excessive heat and rapid expansion of air.

Each of these successive stages of fire is unique and can be detected by fire protection sensing devices that are activated by the characteristics of a specific stage.

Tetrahedron Concept of Fire

The chemical reaction involved in fire is not as simple as the classic fire triangle indicates. A fourth factor, "reaction chain," is of equal importance to the other three elements of heat, fuel, and oxidizing agent. Thus, the fire triangle is converted to a three dimensional pyramid, known as the "tetrahedron of fire." (figure 13-2)

Figure 13-2[3]
Fire Tetrahedron.

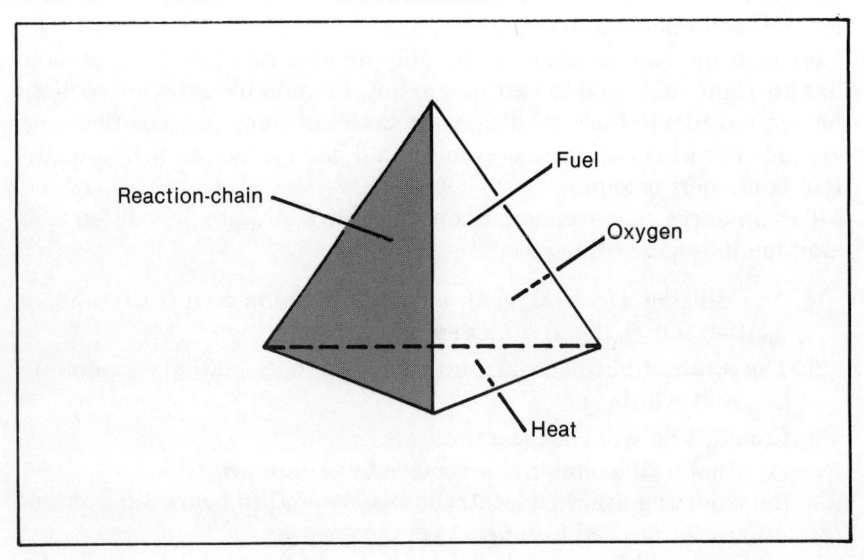

Support for this concept is the discovery of various extinguishing agents that are more effective than those that simply manage to disrupt the triangle. Neutralization of the fire tetrahedron is accomplished by a breaking of the chain reactions by chemical and/or physical means. An example of the process is described by Bush and McLaughlin:

> This reaction-chain is caused by the breakdown and recombination of the molecules that make up a combustible material with the oxygen of the atmosphere. A piece of paper, made up of cellulose molecules, is a good example of a combustible material. Those molecules that are close to the heat source begin to vibrate at an enormously increased rate and, almost instantaneously, begin to break apart. In a series of chemical reactions, these fragments continue to break up, producing free carbon and hydrogen that combine with the oxygen in the air. This combination releases additional energy. Some of the released energy breaks up still more cellulose molecules, releasing more free carbon and hydrogen, which, in turn, combine with more oxygen, releasing more energy, and so on. The flames will continue until fuel is exhausted, oxygen is excluded in some way, heat is dissipated, or the flame reaction-chain is disrupted.[2]

Products of Combustion

The importance of fire prevention and protection can be demonstrated by looking at the products of the combustion process and their effect on life safety. The products of combustion are divided into four categories: (1) fire gases, (2) flame, (3) heat, and (4) smoke. Each of the products can have an injurious, damaging or fatal effect on human life and property.

Fire gases refer to the gaseous products given off during combustion. The chemical composition of these depends on many variables, the principal ones being the chemical composition of the material being burned, the amount of available oxygen, and the temperature or heat present. Many of the gases are dangerous and in sufficient quantity can be fatal. Carbon monoxide, a common fire gas formed when the air supply is very low, poisons by asphyxiation. The hazardous properties of fire gases are particularly apparent in looking at the actual causes of fire deaths. Statistics indicate that fatalities in fire from the inhalation of hot fire gases and hot air are far more common than fire deaths from all other causes combined.

Burns can be caused by direct contact with flame or heat radiated from flames. When flame can be seen it can be assumed that it is rarely separated by an appreciable distance from the burning materials. Flame is a distinct indicator that materials are burning in the presence of an oxygen-rich atmosphere.

Heat is the final stage of the combustion process. Exposure to heated air can cause dehydration, heat exhaustion, blockage of the respiratory tract by fluids, and burns. Burns are commonly classified into three categories: first, second, and third degree.

A first degree burn is characterized by heat, pain and reddening of the burned surface, but does not exhibit blistering or chapping of tissues. A second degree burn is marked by pain, blistering and superficial destruction of the dermis, with fluid accumulation and possible swelling of the tissues beneath the burn. The most serious type, the third degree burn, is characterized by destruction of the skin through the depth of the dermis and possibly into underlying tissues, accompanied by loss of body fluids and sometimes shock.

Smoke, the solid and liquid particles in suspension in the gases, can also have harmful effects. Smoke particles can be irritating when inhaled, and extended exposure to them can cause damage to the respiratory tract. Smoke, with its ability to obscure visibility and inhibit breathing, is a principal life hazard in a fire. Smoke can provide an early warning of fire, but at the same time generate or contribute to panic on the part of observers.

The best protection against fire is prevention. Participants in any situation must, in the interest of life safety and assets protection, understand the characteristics and nature of fire in order to take those steps necessary to prevent the creation of a fire hazard. No environment can be absolutely free of a potential fire. Buildings are never fireproof; however, many do have fire-resistant properties. Participants in any situation are not always safety conscious. Mistakes can be made or unforeseen conditions can develop which make it imperative to always have a working, effective program of fire prevention and control.

Security's Role

Though the organizational placement and role of security may vary from company to company, security will generally have either a direct or indirect responsibility for the development, operation, and enforcement of any fire prevention program. Security operations to protect property against fire usually fall into three categories: (1) to facilitate and control the movement of persons within the premises; (2) to ensure orderly conduct on the property; and (3) to protect life and property at all times. These security operations can include the following duties:

a) Prevent entry of unauthorized persons who might, intentionally or unintentionally, set a fire.
b) Control the activities of people authorized to be on the property, but

who may not be aware of procedures established for the prevention of fire.

c) Control pedestrian and vehicular traffic during exit drills, and control evacuation of the property or parts of it during emergencies.

d) Control of gates and vehicular traffic to facilitate access to the property by the public fire department, members of any private fire brigade, and off-duty management personnel in case of fire and emergencies.

e) Checking permits for hot work, including cutting and welding, and where necessary, standing by to operate fire extinguishing equipment.

f) Detecting conditions likely to cause a fire, such as leaks, spills and faulty equipment.

g) Detecting conditions likely to reduce the effectiveness with which a fire may be controlled, such as portable fire extinguishers out of place, sprinkler valves closed and water supplies impaired.

h) Performing operations to ensure that fire equipment will function effectively. These may include testing automatic sprinkler or other fixed fire protection systems, fire pumps and other equipment related to these systems and assisting in maintenance of this equipment, checking portable fire extinguishers and fire hoses and assisting in pressure tests and maintenance service on these items, testing fire alarm equipment by actuating transmitting devices, and checking equipment provided on any motorized fire apparatus and making the periodic tests and maintenance operations required for it.

i) Promptly discovering a fire and calling the public fire department (and the fire brigade of the property, where there is one).

j) Operating fire control equipment after giving the alarm and before the response of other persons to the alarm.

k) Monitoring signals due to the operation of protective signaling systems provided, such as alarms from manual fire alarm boxes on a system private to the property, signals for water flow in sprinkler systems, signals from systems for detecting fires and abnormal conditions, including trouble signals.

l) Making patrols over routes chosen to ensure surveillance of all the property at appropriate intervals.

m) Starting up and shutting down certain equipment when there is no other personnel provided for the purpose.[4]

Every time a security guard goes on routine patrol or inspection his instructions should include such matters as:

a) Outside doors and gates should be closed and locked; windows, skylights, fire doors, and fire shutters should be closed.
b) All oily waste, rags, paint residue, rubbish, and similar items should be removed from buildings or placed in approved containers.
c) All fire apparatus should be in place and not obstructed.
d) Aisles should be clear.
e) Motors or machines carelessly left running should be shut off and reported.
f) All offices, conference rooms, and smoking areas should be checked for carelessly discarded smoking materials.
g) All gas and electric heaters, coal and oil stoves, and other heating devices on the premises should be checked.
h) All hazardous manufacturing processes should be left in a safe condition. The temperature of dryers, annealing furnaces, and similar equipment which continues to operate during the night, on holidays and weekends, should be noted on all rounds.
i) Hazardous materials, such as gasoline, rubber cement should be kept in proper containers or removed from buildings.
j) All sprinkler valves should be open with gauges indicating proper pressures. If they are closed, this should be reported immediately.
k) All rooms should be checked during cold weather to determine if they are heated properly.
l) All water faucets and air valves found leaking should be closed. If the guard is unable to stop leaks, the condition should be reported.
m) Particular attention should be given to new construction or alterations which may be under way.[5]

Training for Fire Prevention and Protection

Security personnel must be thoroughly acquainted with the property being protected. They must be familiar with the physical features of the property, the materials utilized and stored on the premises, the fire characteristics of buildings, equipment, and materials, the location and operation of fire fighting equipment, and the location of critical control valves and switches. Areas of basic training for security personnel should include the classes of fire, first aid, extinguishing agents, extinguishing equipment and support services.

Hazardous Materials

Central to any training for fire prevention and control is an understanding of fire hazards. A grouping of materials referred to as fire hazards would include:

1) Light combustible materials: wood shavings, cotton, paper, etc.
2) Combustible dusts
3) Flammable and combustible liquids
4) Flammable gases
5) Materials subject to spontaneous heating and ignition
6) Explosive materials, acids, and oxidizing agents.

All materials thus identified as being hazardous should be visibly marked and rated. Marked and rated materials should be categorized as being hazardous in one or more of the following ways: hazardous to health, flammability and/or reactivity. The following system of identifying, rating and categorizing fire hazards has been devised by the National Fire Protection Association.

Colors are used to designate the type of hazard and numbers to denote the level of severity. Figure 13-3 shows a sample sign. In the form of an adhesive label or tag, it can be attached to containers holding materials having hazardous properties, or in the case of a hazardous area can be placed on doors, walls, partitions, etc.

Figure 13-3
Hazardous Materials Marking System Devised by the National Fire Protection Association.

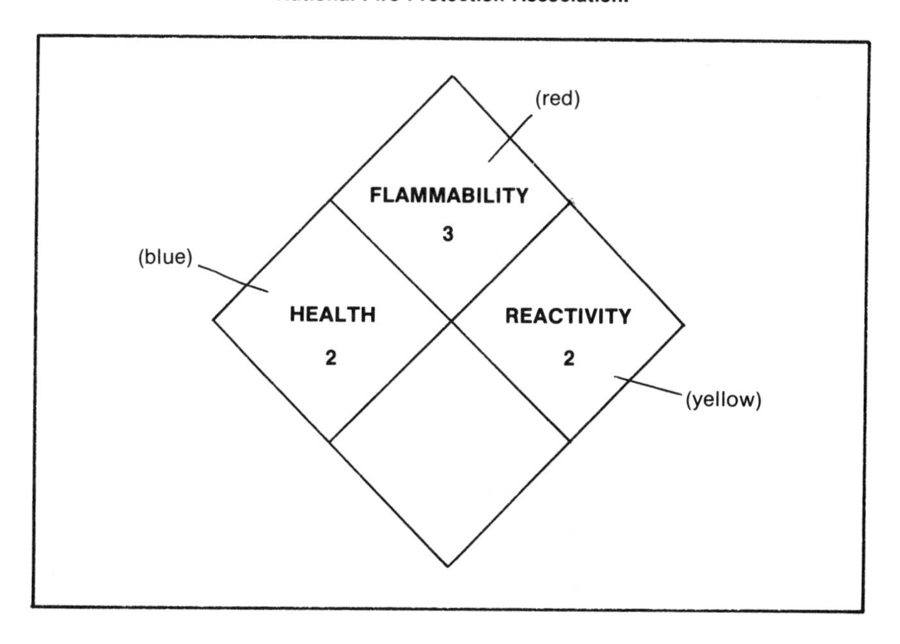

The colors red, blue, and yellow denote flammability, health, and reactivity hazards, respectively. The numbers indicate the order of severity, ranging from "4" indicating a severe hazard, to "0" indicating no special hazard. The fourth and open space can be used to denote any unusual characteristic of a hazardous material, such as radioactivity or reactivity with water. Security should be instrumental in evaluating, inspecting and enforcing any program designed to identify and control the usage and availability of hazardous materials.

Classes of Fire

There are four classes of fire with which security personnel should be familiar.

Class A—Fires involving common combustible material such as wood, paper, cloth, rubber, and some plastics.

Class B—Fires involving flammable liquids, such as petroleum-based gas and oil products.

Class C—Fires in electrical equipment.

Class D—Fires involving combustible metals such as sodium, magnesium, and potassium.

Each of the above classes of fire involve specific materials (fuels), all of which may be present at a given facility. It is essential for security personnel to be aware of and able to classify all the materials used and stored at their facility. A basic knowledge of the combustibles involved in each class of fire is essential for the selection of the proper extinguishing agent.

Classes of Extinguishers

Portable fire extinguishers are designed to discharge a contained amount of fire extinguishing agent. Such a unit can be carried or moved to the scene of a fire. Its effectiveness depends on the use of the proper extinguisher and extinguishing agent for the fire encountered, the proper method of use, the adequacy of the amount of extinguishing agent, and the proper functioning of the unit. Some portable extinguishers are effective on only one type of fire, while others are suitable for two or more classes of fire. Extinguishers are classified as A, B, C or D according to the class of fire they are designed to extinguish.

Class A—Class A extinguishers are used to extinguish fires involving common combustibles which require an extinguishing agent that is heat absorbing. Class A extinguishers accomplish this cooling effect with water, water solutions, or the coating effects of certain dry chemicals.

Class B—Class B extinguishers are used on fires involving flammable liquids. Class B extinguishers smother fires by cutting off the oxygen supply. A number of powdered chemicals, foam compounds, heavy non-combustible gases, and other agents will accomplish this.

Class C—A fire involving electrical equipment should be extinguished by an agent which is a nonconductor of electricity. Carbon dioxide (CO_2) is a common Class C fire extinguisher.

Class D—A fire involving combustible metals requires a heat-absorbing extinguishing agent. Class D extinguishers may contain various dry powders that will not react with the burning metal.

Most manufacturers of extinguishing equipment utilize distinctive markings for extinguishers which indicate the class of fire on which they should be used (figure 13-4).

Figure 13-4
Classes of Fire Extinguishers.

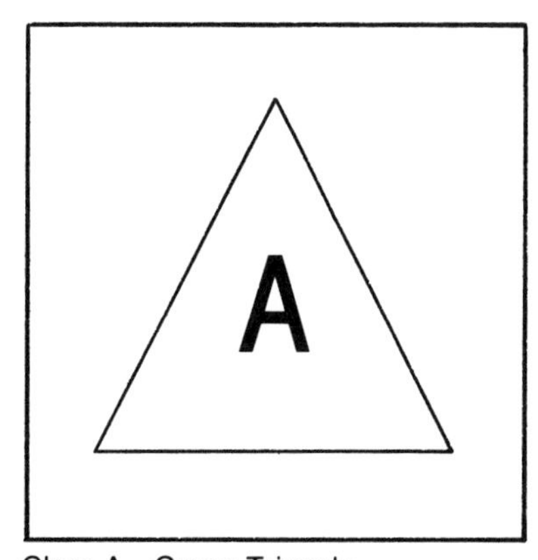

Class A - Green Triangle

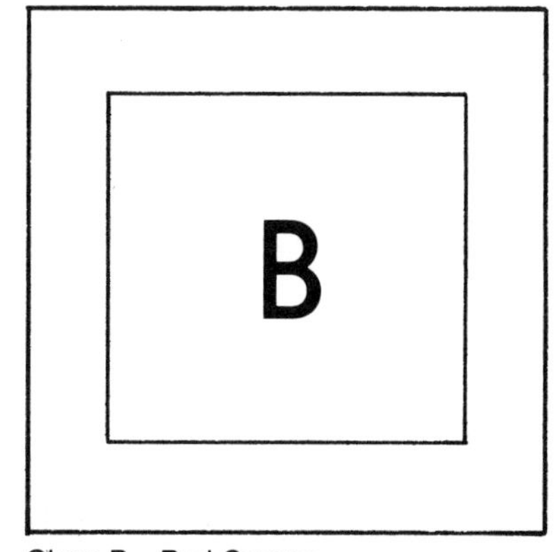

Class B - Red Square

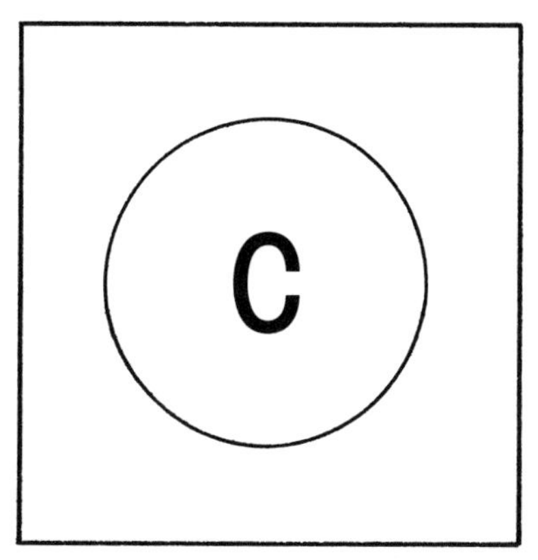

Class C - Blue Circle

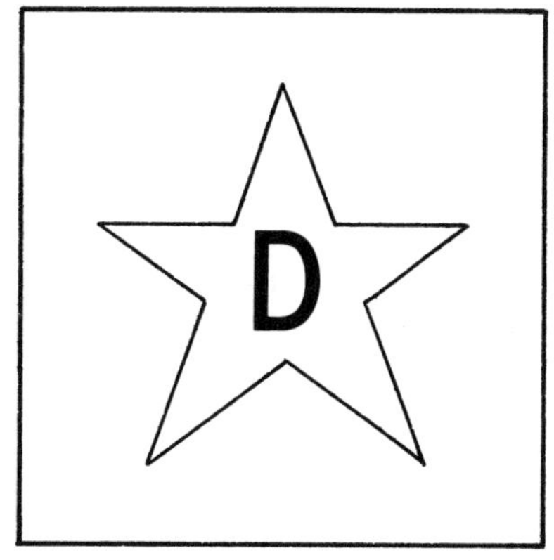

Class D - Yellow Star

Some portable fire extinguishers are rated as multi-purpose extinguishers. For example, some dry chemical extinguishers are rated as ABC extinguishers. However, an extinguisher thus rated may not be as effective on one class of fire as it is on another. The size of the extinguisher, the type of extinguishing agent, and the type of fire determine the rating, i.e., the relative extinguishing effectiveness of the unit. The relative effectiveness of a portable extinguisher is denoted by a number which precedes the class letter on the identifying label. For example, an extinguisher may be rated 4-A: 16-B: C. This indicates that the extinguisher should:

1) Extinguish approximately twice as large a Class A fire as a 2A extinguisher. (A 2½ gallon water extinguisher is rated 2-A.)
2) Extinguish about 16 times as large a Class B fire as a 1-B extinguisher (tested to extinguish 16 square feet flammable liquid pan fire).
3) Be suitable for use on live electrical equipment. (A 1-C rating is the same as a C rating.)

Selecting the proper extinguisher depends primarily on the hazards present at a particular facility. Before deciding which extinguishers to purchase, an in-depth survey of the facility must be made. The survey should give an accurate picture of the potential hazards of operation confronting each specific area and the facility as a whole. Selection of extinguishers must also consider the workforce, i.e., women, for example, may not be able to handle and operate heavy extinguishers, thus, lighter weight units should be chosen when applicable. In the choice of extinguishers and extinguishing agents, the safety of the operator must also be considered because some extinguishing agents when used in closed or confined spaces may react with hot metal or other substances to produce toxic vapors.

After a portable extinguisher is properly selected, security personnel and other employees must be trained as to its most efficient and effective operation. Many companies require that all employees be trained in the use of extinguishers during their initial employment and also receive periodic refresher training. If a proprietary fire department or permanent fire brigade is not present, the responsibility for such training is often given to security.

Another necessary and important aspect in the proper utilization of portable extinguishers is that they must be properly installed and maintained. This means that extinguishers must be located and installed to conform with OSHA standards and other fire regulations. They must be mounted at the proper height or placed in approved containers. Regular inspections (at least monthly) should be made to make certain that extinguishers are where they should be, have not been activated, damaged or tampered with, and are not blocked by materials or equipment.

Extinguishing Agents

The suppression and extinguishment of fire involves two essential elements: an extinguishing agent and a means or system for applying the extinguishing agent. The utilization of an extinguishing agent and its method of delivery must be related to the particular space and property to be protected. The effectiveness of fire suppression efforts depends on the proper selection and application of extinguishing agents if the fire cycle is to be disrupted.

Water

For most common combustibles (Class A) such as wood, paper, and cloth, the simplest and most effective means of removing the heat of a fire is through the application of water. The application of water can be varied and will depend on the properties of the fire. Applying water to a burning fuel, whether by a stationary sprinkler system or portable extinguisher, cools the fuel to the point where insufficient heat is present to support continued combustion.

Foam

Foam extinguishing systems have been used extensively for many years, especially in the petro-chemical industry and crash-rescue units at airports. Foam extinguishing agents, produced by chemical or mechanical means, are air-water emulsions that serve to exclude oxygen from the surface of the burning material. Both mechanical and chemical types of foam are effective on Class A and Class C fires.

Carbon Dioxide

Carbon dioxide is a colorless, odorless, inert, and electrically nonconductive agent that is approximately fifty percent heavier than air. It serves to extinguish a fire by displacing the normal atmosphere, that is, it reduces the oxygen content below the level required for continued combustion. Because carbon dioxide is discharged in a gaseous form by internal storage pressure, and the vapors are heavier than air, carbon dioxide extinguishing systems are usually found in interior locations protecting areas containing electrical hazards, gaseous and liquid flammable materials, and food preparation operations. Portable carbon dioxide units are particularly applicable to electrical substations, motors, computer equipment, office equipment, vaults, etc.

Halogenated Agents

Halogenated extinguishing agents, commonly referred to as halons, are either vaporizing liquids or liquified gases capable of extinguishing and suppressing fires in various materials when applied at proper rates and in proper concentrations. Halons can be used for portable extinguishing units or stationary systems.

A halon is a hydrocarbon in which some of the hydrogen atoms have been replaced by such elements as bromine, chlorine, fluorine, iodine, etc., or by combinations of these in order to create a fire extinguishing gas or liquid. In vapor form, these halogenated hydrocarbons change to other chemical compounds that inhibit or prevent oxidation; however, a number of "halons" are toxic, thus, hazardous to humans. Halon systems must be designed to provide for controlled concentrations when activated, so that occupant evacuations can be accomplished before a hazardous concentration level is reached. Both Halon 1211 and 1301 are widely used for protection of electrical equipment, airplane engines, and computer rooms.

Dry Chemical

Dry chemical extinguishing agents consist of finely divided powders that effectively snuff out a fire when applied by portable extinguishers, hose lines, or stationary systems. Dry chemical is particularly effective for fires in flammable liquids and in certain types of ordinary combustibles and electrical equipment, depending on the type of chemical agent used. The basic chemical agents used include: (1) sodium bicarbonate, (2) potassium bicarbonate, (3) potassium chloride, (4) ureapotassium bicarbonate, and (5) monoammonium phosphate. Dry chemical agents are non-toxic, stable at temperature extremes, and nonconductive.

Dry Powder

The increased use of combustible metals such as magnesium, sodium, lithium, etc., created a need for a special agent to extinguish fires involving these materials. Extinguishing agents developed for combustible metals are designated as dry powders. Delivery systems include portable extinguishers, wheel units, and stationary units. Class D fires create unique problems, and successful control and extinguishment of such fires depends on adequate knowledge and training for the fire fighter.

Sprinkler Protection

Portable fire extinguishers are designed to be taken to the fire location, whereas fixed or stationary fire extinguishing equipment is designed to control and extinguish a fire in a given area. Fixed fire extinguishing systems can utilize a variety of extinguishing agents. They are automatic in operation, designed to discharge the extinguishing agent upon activation.

While various extinguishing agents are available for use in fixed extinguishing systems, the most common type is the automatic water sprinkler system. Standard sprinkler installations usually consist of a combination of water discharge devices (sprinklers); one or more sources of water under pressure; water-flow controlling devices (valves); distribution piping to supply the water to the discharge devices; and auxiliary equipment, such as alarms and supervisory devices. Outdoor fire hydrants, indoor hose standpipes, and hand hose connections are also common components of these systems. The fundamentals of water sprinkler protection evolve around the principle of the automatic discharge of water, in sufficient density, to control or extinguish a fire in its initial stages.

There are two major types of automatic water sprinkler systems: the dry pipe system and the wet pipe system. Which type of sprinkler system is most appropriate in a given location is usually dependent on whether or not the piping would be subject to freezing temperatures as well as the type of materials being protected.

Wet Pipe Sprinkler

A wet pipe sprinkler system consists of automatic sprinklers attached to a system of piping that holds water and is connected to a water supply. Usually the sprinkler head contains a fusible element which will melt at a predetermined temperature and discharge the water, extinguishing fire in the area of the activated unit. Wet pipe sprinkler systems are found in those areas not subject to freezing and where hazardous conditions are minimal. The principal advantage of the wet pipe system is that, when activated, it will discharge water only in the immediate area of the fire. Its primary disadvantage is that the fusible element contained in the sprinkler head must be melted before water is discharged; thus, there is a time lag during which the temperature must build up to the predetermined level of heat necessary to melt the fusible element. A fire developing on the outer fringes or in the area of overlap between two sprinkler heads has an opportunity to grow before the sprinkler head(s) are activated.

Dry Pipe Sprinkler

The dry pipe sprinkler system is used in areas that cannot be properly heated or where conditions of occupancy or special hazards require immediate application of water over a given area. Dry pipe sprinkler systems consist of sprinkler units attached to piping that contains air under pressure. When air is released by activation of a sensing device, this opens a dry pipe valve which allows water to flow into the piping system and be discharged through the open sprinklers. Thus, until a sensing device is activated, there is no water in the piping. A variation of the dry pipe system is the deluge system, which consists of open sprinkler units attached to larger piping, to allow a large quantity of water flow. When activated, a deluge system can release huge quantities of water very quickly.

A routine program of inspection and maintenance of sprinkler systems should be developed. Security personnel are often given the responsibility of inspecting and ensuring that all components of the fire protection system are functional and in good working condition. Therefore, adequate training in the operation and care of sprinkler equipment should be given to security personnel.

Fire Protection Signals

Protective signaling devices play a major role in fire detection and protection. The signaling device may be as simple as a hand operated gong or as complex as an extensive electrical system that covers an entire factory. The device may be a local alarm that can be operated manually or automatically, or one that generates a local alarm and also transmits a warning signal to a remote location. Protective signaling devices can be used for many purposes:

1) to notify people of a fire
2) to call or alert the fire department, private fire brigade, security department, or other receiving locations
3) to monitor extinguishing systems and warn of activation or nonfunctioning condtions
4) to monitor industrial processes and warn of hazardous conditions
5) to supervise people electronically
6) to activate control equipment[6]

Fire protection signaling devices are an integral part of any fire protection system. Because the first few seconds of a fire are so critical in fire protection and suppression, the presence and proper functioning of a protective signaling system may make the difference between life or death. Various types of detection and monitoring devices can be utilized in a protective signaling system. Some are used to monitor water and/or air pressure of the system, and activate when a sprinkler system, fire pump, or other fire protection device is activated or malfunctions. Others detect and react to fire or combustible conditions with or without the activation of fire extinguishing equipment. Some of these include fixed or rate-of-rise heat sensitive devices, ionization detectors sensitive to smoke and gaseous products of combustion, photo-electric detectors sensitive to smoke, pneumatic tube type detectors sensitive to air pressure variation, and so on. (See Chapter 6.)

Safety

Federal, state and local laws and the standards set by insurance carriers require that almost every employer provide a safe and healthful work environment for employees. The Occupational Safety and Health Act (OSHA) passed by Congress in 1970 spells out in great detail what the employer must do to ensure safe working conditions for employees. It specifies requirements for general housekeeping, equipment and equipment operation, environmental health controls, production operations, hazardous material handling and storage, fire protection equipment, materials handling and storage, and so on.

In addition to OSHA, most employers must satisfy state and local regulations relating to safety and fire protection. In fact, some states have safe work laws and environmental protection laws which go beyond those set by OSHA and the federal government. Other restrictions and standards may be set by the insurance companies for their commercial and industrial clients. Many times insurance carriers refuse to provide coverage for potential clients or drop old clients because of unsafe conditions at a facility.

Every organization is dependent for its success on preserving the efficiency and effectiveness of its productive capability. Whether its product is a manufactured item or a service, an organization is dependent on its primary asset, i.e., the people that make its existence possible. To expose this asset to a hazardous or unsafe environment is to invite disruption and even destruction. Safety and loss prevention are prime responsibilities of operating management. An unsafe environment or accident is a sign of something wrong in the management system. Companies are thus motivated to have a safe work environment by moral, legal, and self-interest considerations. The degree of their concern for the safety of employees can range from a "ho-hum" attitude to one of strict, fair regulation. If company leadership projects a strong positive attitude toward safety, it is likely that this same attitude will filter down and throughout the company hierarchy, resulting in high morale and efficient, maximal production.

OSHA

The Occupational Safety and Health Act was signed into law on December 29, 1970. Its purpose was to ensure, so far as possible, safe and healthful working conditions for every working man and woman in the nation. Every business with one or more employees which is affected by interstate commerce is covered by the law.

Under the act, each employer:

1) Has the general duty to furnish each of his employees employment and places of employment which are free from recognized hazards that are causing or likely to cause death or serious physical harm.
2) Has the specific duty of complying with safety and health standards promulgated under the Act.[7]

Each employee, in turn, has the obligation to comply with the safety and health standards, and all rules, regulations, and orders which are applicable to his own actions and conduct on the job. Employers are generally required to post OSHA-required information materials in a conspicuous manner, to keep records and report certain occurrences relative to safety and accidents, to report variances from standards, to cooperate with OSHA compliance officers, and to comply with recommended or mandated changes under the Act. Today, almost all organizations accept the idea of moral, legal, and financial responsibility for work-related injuries.

Workers' Compensation

Workers' compensation statutes are intended to provide compensation benefits for personal injury caused by accident or conditions arising out of and in the course of employment. Worker's compensation programs generally have the following common objectives:

1) Income replacement
2) Restoration of earning capacity and return to productive employment
3) Industrial accident prevention and reduction
4) Proper allocation of costs
5) Achievement of all objectives in the most efficient manner possible.[8]

Thus, workers' compensation acts and programs in the United States provide incentives to employers to introduce measures that will decrease the frequency and severity of accidents. That is, the costs of workers' compensation insurance, which the employer must pay, are tied to safety practices: the safer the work place, the lower the costs.

Accident Prevention

An effective accident prevention and occupational safety program in the work place must be related to proper job performance. When people are trained to function properly in their work environment they will perform more safely.

Safety is a term encompassing the areas of fire safety, personal security, and accident prevention. All employees must be initially trained and periodically retrained in general safety practices and in specific job related responsibilities. Safety awareness, from the top of the management hierarchy to the lowest job classification, is imperative if safety hazards and violations are to be found, understood, and eliminated.

A security department within an industry, institution, business, etc., has a natural role to play in safety and accident prevention. Security personnel who are aware and knowledgeable of required safety standards and practices can be instrumental in correcting or bringing about the correction of safety hazards and violations. The performance of such security duties as patrols, inspections, access control, etc., expose security personnel to most if not all of the work environment, employees, and operations present. It would be a waste if their observations, contacts, and suggestions were not utilized.

Summary

Fire is one of the most destructive forces known to man. It can occur as an act of nature, as an accident, or as an act of sabotage. However, regardless of its source, the potential for its occurrence can be reduced with an adequate fire prevention and protection program. This can only be accomplished if the people involved in the program recognize the problems, understand the possible alternatives and arrive at the best solutions. A knowledge of the chemistry and nature of fire, an awareness of the hazardous properties of the facility to be protected, proper selection of the appropriate fire protection equipment, and effective utilization and coordination of equipment, procedures, and personnel must all be present before an environment can be safe and secure.

Under ideal circumstances, the security department would provide support services to specialized fire protection personnel, since fires and emergencies require both security and fire department services. However, the situation often demands that security personnel also have the duty of being fire fighters. Where this is the case, security personnel must be trained to make them better members of a loss prevention team.

Safety is an important aspect of any environment. Legislation such as OSHA and workers' compensation laws has mandated that most work environments be safe and healthful places to work. Of course, the role of security in safety and accident prevention in the workplace, while perhaps a natural addition to security functions and responsibilities, is determined by management.

Discussion Questions

1. What are the factors necessary for combustion?
2. Outline the stages through which a fire usually progresses.
3. What roles can security play in protecting a facility against fire loss?
4. Distinguish between wet pipe sprinklers and dry pipe sprinklers.
5. List and briefly explain the significance of the various groups of materials referred to as fire hazards.

Notes

1. John L. Bryan. *Fire Suppression and Detection Systems*, Glencoe Press, 1974.
2. Loren S. Bush and James H. McLaughlin. *Introduction to Fire Science*, Glencoe Publishing Company, 1979.
3. Bush and McLaughlin, p. 31.
4. National Fire Protection Association. *Fire Protection Handbook*, 1971.
5. *Fire Protection Handbook*.
6. National Fire Protection Association. *Industrial Fire Brigades*, 1968.
7. National Safety Council. *Accident Prevention Manual For Industrial Operations*, 1974.
8. *Accident Prevention Manual*.

Chapter 14
Emergency and Disaster Control

If someone surveyed a small sampling of typical businesses, industries or institutions to determine how they fared during emergency or disaster conditions, a wide range of responses would be collected. Some organizations are well prepared to safeguard the lives of their employees (and anyone else in their facilities), protect their physical plants, and often arrange to continue production (if only on a reduced basis) soon after a crisis. Other organizations may not be as well prepared, but are able to continue functioning due to a unique product or service they provide, or due to their escaping the worst consequences of a disastrous situation. However, the majority of American businesses, industries and institutions suffer catastrophic losses every time a disaster strikes their organization. These different results are not necessarily due to differences in wealth, physical size or number of employees but rather are due to different degrees of emergency and disaster planning.

The following are three brief descriptions of actual disasters and how different organizations coped with them. Attention should be given to the fact that in the two cases where regular security personnel were available to the organizations, the security forces were not given overall control of the response to the emergency, nor did they provide the total emergency response. Emergency operations must be the responsibility of all parts of an organization, and security forces are only a part of the organization.

In a 24-hour period one recent winter, a severe snow storm blanketed a Midwest region the size of Indiana with twelve inches of snow. Electrical power was disrupted over half the area from three to thirty hours. Roads and highways were impassable, because the snow removal equipment could not keep up with the drifting snow. Thousands of persons were stranded away from home or from work, and five people were later found frozen to death in their vehicles. One victim had left his job in a medium-size manufacturing plant when the power was interrupted. The plant did not have an emergency source of power. All employees were dismissed from work; most of them stayed in the plant, while several took rooms at a nearby motel. The power interruption caused irreparable damage to some of the equipment and the mixture of chemical products. After drinking alcohol they had brought into the plant, three employees became involved in an altercation in which one employee was seriously injured. The small cafeteria did not contain enough food to feed all the stranded employees. The marooned workers were restless, worried about their families elsewhere, and became generally unruly. The production line was not restarted until two days after the storm.

This plant had no security force nor emergency plans. With proper planning and preparation the facility could have prevented one death and one incarceration, and saved one day's worth of chemical mixture, the damaged equipment, two work days on the production line and possibly the majority of the over $300,000 in direct losses sustained as a result of this one snow storm.

In the next example, planning and preparation preceded an emergency situation and thereby contributed inestimably to both good public relations and community safety during the week-long emergency.

It began with a radio announcement that a local nuclear power plant had an uncontrolled radioactive leak. Some local residents were advised by radio news personnel to evacuate the area.

The security directors of several large nearby firms and of the local telephone company set their emergency operations into action. They checked various sources for the urgency of the problem and then continued business as usual with their normal operating personnel. Employees were continually apprised of the situation and continued working during their regular shifts. Arrangements were made to have additional contract guards stand by, because of the anticipation of looting should there be a further evacuation of residents. Shoebox-size packages were packed with medication and other supplies needed by individual patients in the event that the local hospital would have to be evacuated. The local police were fully mobilized and the state national guard was placed on alert, ready to move within hours. By the end of the first week most of the residents who had fled the area earlier in the emergency returned, while others decided to leave at that late date.

No one was injured or died in this near-disaster. The industrial firms with emergency preparations continued business as normal with few absentee workers. The emergency was treated as real and lessons were learned to incorporate later in the emergency response plan. Panic, injuries and deaths were avoided by intelligent emergency plans properly employed by trained personnel.

The third example illustrates procedures employed in the aftermath of a disaster. After a recent hurricane caused over 700 million dollars in losses and destroyed electrical transmission to over 300,000 customers in one metropolitan area alone, emergency crews of repairmen were assembled from power companies in adjacent states. The repair process lasted there weeks and was compounded by thousands of downed poles, transformers and power lines that were susceptible to theft due to the high value of scrap aluminum and copper.

Additional contract security guards were hired from outside the affected area because all local guards were already employed to protect other damaged businesses. Fuel, food and other supplies were imported and made available to the power company personnel and guards. The security department had a three-fold objective during this repair phase. One was to patrol the areas with the largest amounts of power company property lying on the ground. Another was to coordinate the deployment of guards to protect repair equipment, supplies and vehicles. The third objective was to protect damaged power company buildings and their contents. This latter objective was made easier by the fact that the power company had already removed some high value equipment to a more secure area in anticipation of damage to the building and subsequent looting. Again, prior planning and preparation for emergencies paid generous dividends by keeping to a minimum further losses of property after the disaster.

In the first example above, the manufacturing plant was not prepared for emergency situations, and when they did occur, they were unnecessarily expensive in the loss of personnel, equipment, products and productivity. Another important loss was the loss of faith in the organization by the employees.

The next two examples demonstrate the successes of organizations in reducing risks to their employees, protecting their property and continuing their work operations. By providing emergency leadership, insuring the safety and welfare of their employees and continuing to serve their communities, these organizations gained not only the esteem of their communities but also the respect and future support of their employees.

The difference between the successful and the unsuccessful responses to the disasters can be directly attributed to careful planning and appropriate preparedness.

The Role of Planning

The only thing certain about planning to protect lives or property from natural or man-made emergencies or disasters is that there is no location, anywhere in the world, that is absolutely free from danger in one form or another. The varieties of potential dangers can be identified, and measures can be taken to reduce the risk of exposure to those dangers by people or property. Planning will not prevent a flood, an explosion, or a strike; however, thorough planning and preparation may prevent the escalation of a danger into a catastrophe.

Unless the organization is very small, the planning to cope with emergencies or disasters should be done by high ranking representatives from all branches, departments or sections, working together in a committee. The planning committee should not be too large or agreements will be difficult to reach, but it should be large enough to ensure representation from all segments of the organization. Someone with knowledge of the security field, whether a staff member who has the responsibility for security or an outside security consultant, should be on the committee. However, the security manager should not be appointed as the committee head or sole investigator, unless he holds a high position of authority in the organization. The following factors should be included in the planning process for an emergency and disaster control plan.

Authority

The owner, manager, or governing body of an organization should prepare a simple, written order to authorize the committee to develop the emergency control plan. The order should provide the committee with the necessary authority to develop a written plan, and then organize, train and assign responsibilities to an emergency force within the organization. This statement should be brief and flexible so that the committee can adjust its deliberations as necessary.

Existing organizational rules and regulations will have to be researched to ensure that there is not already an emergency plan that has been overlooked or forgotten. A previous plan may or may not be adequate for the present situation, so it should be carefully reviewed and updated or declared void. Provisions for the new or revised emergency plan should follow the organization's established rules and regulations, so that there is a formal, legal foundation for the emergency control program.

Written authority should exist to organize certain personnel into a special force during declared emergencies. There may be personnel occupying certain responsible positions who would object to such an extra assignment, or who might be physically or emotionally unsuited for assignment to an emergency force. For this reason, an effort should be made to attract willing and able volunteers from the organization for the emergency force. Consideration for assignment to the force should be given to persons with unusual abilities and special interests that would be beneficial.

The personnel selected for the emergency force should undergo training in primary and alternate emergency duties. Cross-training the emergency force allows for a greater range of assignments and for the assumption of responsibilities in the event that another member is absent or becomes incapacitated.

Within larger organizations there may be authority for personnel to be assigned as emergency coordinators for shifts, floors, buildings, or satellite facilities. This way there will be trained personnel available throughout the organization to provide a continuity of calm leadership in an emergency.

The order of authority should also provide a chain of command that specifies the individuals who have the authority to order certain activities or changes, or assume leadership roles in the event that higher ranking personnel are unavailable during an emergency.

Vulnerability Assessment

Organizations that want to develop an emergency and disaster control plan must be aware of their existing vulnerabilities so that effective safeguards can be planned. A vulnerability assessment is performed in several phases. The first phase involves studying the frequency of the natural or man-made emergencies, the second concerns the physical plants and the third assesses the present guard force.

A small committee should be appointed to conduct the vulnerability assessments. The first assessment merely identifies the types and frequencies of all natural and man-made emergencies and disasters recorded in the past in the area of the organization. If certain types of emergencies recur with regularity, predictions can be made as to the periods of the threats. Natural disasters seem to occur more regularly than man-made disasters.

The second phase of the vulnerability assessment requires an estimation of the structural strength of the buildings of the organization. An engineer should assist the committee in determining which of the buildings are sufficiently strong for use as shelters and which are structurally unsound and unsafe for use during natural disasters.

The third phase of the assessment should be an analysis of the organization's present security measures. The guards may be proprietary or contract and their terms of employment may be such that they would not be available during periods of emergency. The employment of off-duty police officers is an example of this problem: they would be mobilized by the police department during a general emergency and would not be available to the private organization.

Another vulnerability assessment that should be made is the extent to which the organization would be affected by a disruption in electrical, water, sewerage and communications services, and the degree of isolation the facility would experience in event of flooded roadways, collapsed bridges or other obstructions to normal transportation. The assessment should determine what alternate outside sources for these services would be readily available under disaster conditions, and assess those supplies available from within the organization itself.

In those situations where the organization operates largely using outside cartage, supply or labor under normal conditions, the capability of the vendors of those services should also be assessed to determine their ability to perform under certain emergency conditions.

Determining the Required Response or Security Needs

The vulnerability assessments should result in lists of vulnerable points in the physical structures and surrounding areas, the maintenance of power, communications and supplies, and sustained personnel support. Ultimately, the safeguarding of lives is of prime import. Therefore, the disaster plan developing committee should recommend to the facility administrator an order of priority of emergency responses to undertake under various threatening conditions. Questions such as when to order the evacuation of employees to in-house shelters or to other safe locations, whether any personnel should be exposed to danger by remaining in the insecure facility during the disaster, or when to order a shutdown of the production line must be resolved and instructions must be included in the disaster plan.

There must be a preestablished set of conditions and instructions prepared to cover any emergency. There should be limited personal discretion allowed for personnel involved in emergency operations.

Appointment of the Emergency and Disaster Control Coordinator

Large organizations may have the resources to appoint a full-time emergency or disaster control coordinator to manage the rest of the development or implementation of the plan. In smaller organizations someone may have to assume this role part time. In any event, the coordinator should be of sufficient stature and authority in the organization that he will be able to deal effectively with others at all levels in the organizational hierarchy. Not only must he understand the value of preparedness, but he must engender support for the program in periods when emergencies and disasters appear to be no threat. In all cases he must ensure that every facet of the plan be cost-effective.

The appointment of the coordinator does not mean the end of the original committee's responsibilities in further developing the disaster plan. Rather, the appointment will be another sign from top management that there will be support for this project. The continued active use of this committee will also help in gaining support from the separate departments which the members of the committee represent.

Gaining Management Support

The next step in the planning process is to acquaint management at all levels with the vulnerabilities of the organization to emergencies or disasters, and to solicit their advice and cooperation for the development of a disaster control program. The support of management is necessary because there will need to be finanical expenditures, personnel use, and the possible acquisition or sharing of equipment.

Use of In-House Personnel

Regardless of the size of the organization, the presently employed personnel should perform the majority, if not all, of the necessary emergency duties.

The nucleus of the emergency organization can be those personnel already trained and utilized to perform routine and emergency services, such as the foremen, supervisors, and other key administrative personnel, as well as the existing medical, fire, safety and security staff. Other emergency response personnel can be selected from the regular employees who have demonstrated special talents or interests that could be useful, such as CB or ham radio operators, volunteer or auxiliary police or firemen, first aid instructors, or recreational vehicle operators (boats, snowmobiles, aircraft).

The regular maintenance or engineering employees could help to survey, establish and maintain emergency facilities and make damage assessments. Personnel administrators could be responsible for the shelter and welfare services in the facility. Any research or scientifically oriented personnel could be used to monitor radiological sensors or other essential equipment or communications. As the emergency preparedness plan develops, other employees may be found to have special skills or an interest in training to acquire additional skills that would contribute to the response effort.

Sources of Assistance

The previous section outlined the way to utilize in-house personnel as staff for an emergency or disaster situation. Planning must also incorporate as much self-help as possible during such emergencies to minimize the cost to the organization.

Self-help. An important aspect of the vulnerability assessment is to identify the activities and material that are presently located within the surveyed facility. Not only must the regular employees be selected and trained for their responsive roles, but any material or supplies, such as food, water, tools, portable equipment, lumber, or other property that could be used to protect or repair the facility or to recover from interrupted services must be inventoried and kept in useable condition. The vulnerability assessment should also estimate the amount of emergency supplies (sandbags, plywood, nails, etc.) that would be necessary in the event that a certain building or facility had to be protected or occupied throughout an emergency. Many organizations will have within their facilities supplies that could be utilized as makeshift material to block doors and windows, provide a limited water supply, or serve as emergency rations.

In the case of large organizations that are widely dispersed geographically, the assessment might identify such necessary items that would be available within the organization, and the effort that would be required to shift those items to wherever they are most needed. Such organizational self-help is usually less expensive and less objectionable to the organizational managers than would be a request to purchase extra material and equipment that would be of use only during an emergency.

Mutual aid agreements. It is well known that people help each other during emergencies. What is necessary to benefit the most from this trait is to plan ahead for those emergencies that are most likely to occur and thereby maximize the chances that help will be available from and to others as needed.

A mutual aid agreement may involve either distant facilities of one organization, similar organizations, all businesses or activities within a neighborhood, or governmental agencies. There are advantages and disadvantages in the use of mutual aid agreements, so the committee doing the planning will have to explore all possible assistance, select the most advantageous, and maintain open communications for changing to more appropriate sources if necessary.

As mentioned in the section concerning the use of in-house personnel during emergencies or disasters, whenever there is an emergency, certain personnel (national guard personnel, fire fighters, police officers) may have a legal obligation to report elsewhere, and so will be unavailable for emergency or disaster assignments within the organization. When in-house personnel are unavailable and manpower is needed, another organization might have a list of volunteers who would agree to assist the requesting firm. There should be no high expectations that temporary employees can be hired through normal employment agencies during emergencies, because those personnel are less likely to have a sense of loyalty or trust that would be needed in an emergency. Naturally, supplies, equipment, special clothing, food and some financial remuneration must be made available to any outside personnel, just as would be done for in-house employees working beyond normal periods. The legal liabilities for damages, injuries and deaths caused by both in-house and mutual aid agreement personnel must be established during this planning stage.

When some supplies, material, or equipment are not available within the organization, a mutual agreement with the nearest facility that has such items may be warranted. However, it must always be considered that whenever one facility is in need of such items, the same items may also be needed by the facility that possesses them. Another factor to be considered is that when there is a dire need for certain property that is in scarce supply, the price may become very high if not exorbitant. For these reasons, alternate sources should be identified locally, as well as from facilities that are some distance away from the affected area.

There are some mutual aid agreements between private firms and governmental agencies, such as fire or police departments. In such agreements private firefighting equipment and personnel are pledged to stand by whenever the public firefighters are called out. Private security personnel are also afforded limited police powers when they are deputized or are provided such limited powers by local ordinance for special occurrences. In such cases the mutual aid agreements are limited, because the police and fire departments may have stronger mutual aid pacts with other governmental agencies that would take precedence.

Government and public services. It may be surprising, but even organizations that prepare emergency plans often do not take advantage of available government information on the subject. The emergency planning committee should contact the various government agencies that can provide this assistance.

a. *The Federal Emergency Management Agency* (2400 M Street N.W., Washington, D.C., 20472) provides published information and recommendations on how to counteract and minimize losses caused by enemy action, natural or man-made disasters. That office should be contacted for an up-to-date bibliography of related U.S. Government Printing Office publications.

b. *Regional emergency and disaster agencies* can provide additional information that may have a more local orientation concerning emergencies that are most likely to occur. As the local coordinating agency for emergency and disaster situations, they also can provide recommendations for local responses to ensure the fullest cooperation and protective measures appropriate to the area.

c. *State and local governments* provide police, fire and other services, including environmental protection, health and welfare and many more services during normal periods as well as during emergency conditions.

d. *Public and public service agencies* may provide water, electricity, telephone communications, sanitation and other services. These agencies should be contacted to learn how they plan to continue their services during an emergency.

Not every police, fire department or other public agency will be able to provide the same level services. For that reason, it is the planning committee's responsibility to establish exactly what services would be available during various emergency situations, and acquire information on how alternative services can be obtained.

One problem seldom considered in emergency planning is the identification of employees with unusual health or disability impairments. There are many people in the general population who can function only with constant medical attention, medication and clinical services. Those persons handicapped by severe color or night blindness, diabetes or epilepsy, for instance, may not only be unable to contribute to emergency functions, but may require special help and attention themselves during the emergency. Therefore, such people should provide instructions as to how they might be helped (name and phone number of physician, medication source, etc.) in the event their malady is aggravated during an emergency.

The vulnerability surveys should have determined the areas with a need for additional personnel support during emergencies. There are several ways to increase personnel strengths from within the organization. One way is to eliminate or decrease non-critical activities and reassign those personnel temporarily. Another way would be to reduce administrative staffs by deferring or postponing non-essential duties until after the emergency subsides. A third way would be to solicit spouses and relatives of present employees to either contribute their services or to accept temporary employment. The employment of additional personnel from among people with a vested interest in the organization may be more productive and involve less risk than hiring outsiders who would have no loyalty for the organization.

Developing the Plan

The existing organizational structure should be utilized as much as practicable for the emergency plan. The regular supervisory authority, the technical skills, and the equipment and material at hand familiar to the employees should form the foundation for emergency actions. In this respect, the disaster organization will not be seen as a separate entity and possibly as a threat by the remainder of the organization. The following factors should be considered in the development of the disaster plan.

1) The conditions requiring the activation of the emergency or disaster control program should be listed.
2) The person or persons with the authority to declare an emergency should be identified.
3) Maps and blueprints of the facility to be protected should be provided.
4) An emergency control center and an alternate site should be designated.

5) Communications facilities between the control center, the major sections of the facility, and the community should be established. Alternate means of communication (radios, sound-powered phones, etc.) should be provided.
6) An emergency organizational hierarchy should be developed and a list of employees and their emergency duties should be available.
7) Emergency shut-down procedures should be developed, and lists made of critical property and records to be secured.
8) Emergency evacuations should be planned.
9) Shelters should be identified and supplied with food, medical supplies, water, and disaster equipment.
10) An augmented security force should be planned.
11) Damage assessment and repair teams should be designated.
12) Emergency power, fuel and utilities should be ready to be activated as needed.
13) Plans for cooperation with federal, state and local police, fire and emergency preparedness officials should be developed. Mutual aid pacts should be set up.
14) Records should be kept of all activities that take place during the emergency for satisfaction of legal liabilities and possible later revision of the disaster plan.

A sound emergency and disaster control plan is the result of thorough advance planning, testing, revision, and updating. An adequate emergency plan must contain detailed and timely information about the resources that can be utilized in the most efficient and expeditious manner.

The plan for emergency control must consider the goals, products, and personnel within the facility. Goal changes, shifting responsibilities and personnel turnover may require the frequent updating of the basic plan. Therefore, emergency planning must be a continuing process if the best safeguards possible are expected.

All planned emergency control measures must be arranged so that they complement and supplement each other, the regular security force and the primary goals of the parent organization. Poorly integrated emergency control measures may result in the waste of manpower, funds, and equipment. Of greater importance, the lack of integration in the plans may jeopardize the safety and security of the facility.

The Role of Security

Security forces have a crucial role during emergency operations. There has been little reference to the regular security forces up to this point because emergency or disaster planning for any type of an organization entails the orderly and efficient transition from normal to emergency operations by the regular work force. The energy and power of the regular work force is many times greater than the security personnel could provide. Still, the role of security is crucial in that a greater degree of protection must be provided in a period of crisis.

The size of the guard force available to an organization during an emergency is usually the same as under normal conditions. There are few proprietary or contract guard services that have large numbers of reserve personnel who wait for an emergency or disaster to happen. Because of the limited numbers of guards that are available to a firm, their roles usually do not differ greatly during normal and emergency operating conditions. Yet because of the aura of authority associated with the guards, most people expect them to be thoroughly trained in every aspect of the emergency operation. This means that the guard forces should know and understand every aspect of the emergency or disaster program, even though their role remains security-oriented. During the emergency they should be kept abreast of programs of the operation so that they can be a source of information for other employees.

Security personnel have the responsibility for ensuring that only authorized personnel are admitted to the facility, that restricted zones are for limited personnel only, and that personnel entering and leaving the plant should be properly accounted for and monitored. The logging of employees and visitors in and out of the facility may seem unnecessary, but during emergencies the log may be of inestimable value in locating persons, in identifying those in certain areas, and possibly for the indentification of bodies.

The guards should be trained to perform security duties during emergencies. They may be given cross-training with fire personnel in order to augment their forces or to learn how fire or safety personnel perform rescue operations. But it must be remembered that rescue operations are best conducted by those forces who are provided with specialized equipment, training, and in-depth knowledge of the unseen dangers of the facility such as high-voltage power lines, deadly fumes, or dangers of explosion.

The guard service must be organized in such a way that they can be deployed to protect the areas of high vulnerability. Temporary measures to secure areas from intrusion by the use of barbed wire, vehicle emplacement, trespass signs, and various alarm systems should be considered. Another important point is to ensure the availability of guard personnel and transportation to respond promptly to calls for assistance. When the facility remains in full or partial operation during the emergency, there are certain areas that will need extra guard help to facilitate the movement and control of employees, such as gates and parking lots. During an emergency, security duties will increase beyond the capability of the available personnel, so plans should be made for the augmenting of the security force with persons selected from the regular work force. There may be employees with previous police or military experience who would require little training or supervision to provide needed support services to security efforts. When provided with a special armband or stenciled hard hat, they might be employed at gates to check identification or as the second member of a guard patrol.

Finally, the question arises about arming the security personnel. Many security forces are not armed. Some have arms available to them via their guard supervisors or in arms lockers. Others are armed at all times. The value of equipment or merchandise, the danger to life or property, recent criminal attacks in the area and other factors will dictate the armaments necessary. The emergency and disaster control plan committee should resolve that question in consultation with the organization's legal officer, the police and the prosecutor's office.

Training for Emergencies

The objective of a training program is to ensure that all the emergency force personnel are able to perform their special duties quickly and efficiently. The extent and type of training required to properly prepare emergency control forces will vary according to the importance, vulnerability, size and other unique factors affecting the particular facility.

Thorough and continuing training is the most effective means to obtain and maintain the maximum proficiency of the emergency force personnel. Regardless of how carefully emergency personnel are selected, it is seldom that they will initially have all the qualifications and experience to be effective team members without cooperative training. Furthermore, new and revised emergency requirements frequently mean that the personnel must be retrained. The gap that exists between the job requirement and personal ability can be bridged by training.

Not all of the personnel selected for the emergency force will have the same training needs. It would be a waste of valuable time to require all personnel to sit through the training for the other participants when that training is specialized or has little relationship to everyone's assignments. Thus the past experience, education, training, acquired skills and interests of each emergency force member should be taken into consideration.

Thorough training has many benefits for both the organization and the emergency control program. Supervisors realize benefits because the trained personnel are easier to supervise: there is less wasted time and fewer mistakes are made. The resultant economies are of benefit to the organization. The confidence that is instilled by training affects the morale and welfare of the entire organization.

Individual emergency personnel also benefit from training through increased knowledge and skills, through increased opportunities for personal advancement and through a better understanding of their relationship to the emergency program and the parent organization.

Testing the Plan

When the disaster plan has been completed, provisions for its implementation and testing must be made. This is the time when deficiencies and unrealistic features of the plan are discovered and corrected.

The main goals of such a test include the following:

1) To familiarize all personnel in the emergency force with the overall plan and to acquaint each person with his own emergency duties and responsibilities.
2) To evaluate the workability of the plan and identify deficiencies.
3) To make the necessary adjustments to the plan for future testing and implementation.

There are different ways to test an emergency disaster control plan. The most comprehensive method would be to test the plan in three phases. The first phase would be to test the key individuals separately. Then the plan would be tested in each section or department within the facility. The third phase would be a test of the response of the entire facility. Various conditions could be simulated for each phase, or if one threat is considered more likely than the others, one type of emergency could be simulated for all three phases.

After the emergency control coordinator and the planning committee are satisfied that they have a workable plan, they might coordinate further testing with local emergency preparedness exercises to determine the plan's compatibility with the local effort.

There is no hard and fast rule as to how often an emergency plan should be tested. There must be a continuing and concerted effort to upgrade the response capabilities of the personnel selected for the emergency operation. In order to keep personnel familiar with the plan and to familiarize new personnel with their responsibilities, further individual, departmental and facility testing should be conducted as needed, at least annually and possibly semi-annually.

Mock Drills. After the individual departmental and facility training has been determined to be satisfactory by means of the earlier described testing, the periodic, semi-annual or annual testing could take place as mock drills. The drills should be arranged to simulate various types of natural or man-made emergencies. Realism may be achieved by preparing appropriate simulations of damage to property, injury to role players and deaths. Prior knowledge of the specific target of the mock drill should not be disseminated to the participants. They will then be required to "solve" their problems spontaneously, based upon their training.

Those conducting the mock drill should delimit the target area to one building or area in such a way as to interfere with normal organizational operations only minimally. Markers may be placed around the target area denoting damages and hazards. Role players may be provided with artificial replicas of injuries or can have parts of their bodies painted with food coloring to resemble blood. Other role players can represent dead or dying victims.

The emergency control personnel and their supervisors would be provided with an instruction booklet that would describe individual problems that each participant would be required to solve.

Persons not directly involved in the emergency control program (and possibly not even organizational employees) should be employed to act as judges to assess the appropriateness of the responses made by the participants.

The mock drill must be long enough for the participants to be informed of their roles, gather the necessary equipment, and make judgments about the simulated damages and injuries, and take the appropriate action. The mock drill should culminate in a session where the participants and judges evaluate the actions taken and observed.

Discussion Questions

1. Discuss the major aspects of planning for emergency situations and how they interrelate.
2. What specific written authority should the emergency plan include in order to direct the emergency force operations and personnel?
3. Identify the major phase of a vulnerability assessment for emergency or disaster operations.
4. Discuss the advantages and disadvantages of limiting the discretion of responses taken by personnel in the emergency force.
5. Identify some of the problems that will limit the use of all in-house personnel during emergencies.
6. Discuss the advantages and disadvantages of mutual aid agreements for assistance during emergencies.
7. List several benefits associated with training the emergency force.
8. List the different ways an emergency or disaster plan can be tested.

Chapter 15
Private Security Education

Security Education

The past and present efforts to provide quality security education can best be described as limited. However, there are some encouraging signs.

The first significant efforts in educating persons for private security occurred in the late 1950's at Michigan State University, with the establishment of a Bachelor of Science degree in Industrial Security Administration within the School of Police Administration. The first degrees were awarded during the 1958-59 academic year. It is significant that Michigan State University was already a recognized leader in police education and the degree program was connected with that effort.

Most private security degree programs since this initial effort are also located within the broader program areas of criminal justice. There are, however, significant numbers of private security professionals who feel that this was the wrong approach. They feel that private security education programs should be connected with business programs. A close look at most private security education programs indicates that the developers of the programs are aware of this philosophy since, almost without exception, the curricula include required and/or elective courses in business, as well as broad general education courses, criminal justice courses, and the specific security courses.

One of the most influential publications relating to the field of private security is a publication distributed in 1972 from the American Society for Industrial Security/ASIS Foundation, Inc., entitled *Academic Guidelines for Security and Loss Prevention Programs in Community and Junior Colleges*. Most of the certificate and two-year associate degree programs use the curricula guidelines in this publication. Later in this chapter the guidelines are reproduced with a further discussion on associate degree programs.

From an employment standpoint, the opportunities are most encouraging. Assistant Professor Robert Fischer, Department of Law Enforcement Administration, Western Illinois University, surveyed 1,172 WIU law enforcement graduates in 1979 and found that 15.2 percent were employed in the private security field, even though WIU offers only a minor in security administration. On a raw number and percentage basis, only public law enforcement had more persons employed. Other components of the criminal justice system (such as courts and corrections) had less.

Another survey of Western Illinois University graduates was conducted in 1985. By the time of the survey, Western had over 2,300 graduates. Once again private security, at either the entry or administrative level, was second only to law enforcement careers. Thirty-eight percent were in law enforcement and ten percent in security. Other career fields, such as corrections, probation and parolee, had much smaller percentages.

As part of his doctoral dissertation, Clifford W. Van Meter, Director, Police Training Institute, University of Illinois, also surveyed Western graduates from 1971 to 1976. This research indicated that approximately ten percent of the graduates were employed in private security.

Additional research conducted by Dr. Robert J. Fischer and reported in his doctoral dissertation provides additional information regarding employment in private security. Most graduates enter employment as management trainees in retail and industrial organizations. This is primarily because the starting salaries are reasonable and opportunities for promotion, especially in the retail field, attract graduates of private security education programs. Similar salaries and opportunities are not prevalent in contract security, which tends to reduce the number of college graduates seeking employment in contract security.

Although there has not been a rapid increase during the last decade in the percentages of graduates entering private security, the entry has been steady and can reasonably be expected to increase as the private security industry continues to grow — especially in relation to public law enforcement employment opportunities.

Guidelines for Degree Programs for Private Security

In 1977 the Report of the *Task Force on Private Security* provided an excellent overview of the entire subject of private security education. The Commentary for Standard 8.4 of the report is included because it provides a historical perspective and most of the observations and recommendations are still pertinent at the present time.

Standard 8.4

Degree Programs for Private Security

The private security industry and the Law Enforcement Assistance Administration (LEAA) should cooperate in the encouragement and development of:

1. Certificate, associate of art, or associate of science degree programs designed to meet local industry needs;
2. Undergraduate and graduate programs designed to meet private security needs.

Commentary

Although there is presently no comprehensive involvement by colleges and universities to provide educational opportunities for private security personnel, it should also be recognized that there is little evidence that the security industry or government agencies have encouraged their development. This standard is based on the premise that the industry, LEAA, and educational institutions can cooperate for mutual benefit.

Certificate and associate degree programs designed to meet the needs of the private security industry are a recent, but potentially significant, resource for improving the delivery of security services. *Academic Guidelines for Security and Loss Prevention Programs in Community and Junior Colleges*, published in 1972, identified 5 certificate programs, 2 associate programs, and 58 junior or community colleges offering at least one security course. Research revealed 6 certificate programs, 22 associate programs, and 49 junior or community colleges offering at least one security course. The number of junior and community colleges offering some form of private security education grew from 65, in 1972, to 77, in 1976. However, a closer look beyond these positive indicators of the growth of private security education reveals a need for much greater effort. Only five states (California, Illinois, Michigan, New York, and Virginia) have five or more programs at the junior and community college level; 24 states do not have even one institution that offers one course. Thus, although there has been growth in educational programs, the future offers great challenges to junior and community colleges to help develop the skills, knowledge, and judgment needed by private security personnel through appropriate courses. (See Figure 15.2 and Appendix C for updated information.)

Certain critics have voiced the opinion that because degree programs in business administration, criminal justice, law enforcement, and other related fields have provided appropriate educational backgrounds in the past to persons in private security, there is no need, at this time, for specific private security degree programs. This position is in error. Private security degree programs will not only enhance the professional movement in private security but also promote needed research and technological advancements.

Three significant resolutions passed at the First National Conference on Private Security are pertinent to the future development of educational programs. These resolutions are:

1. A multi-disciplinary and scholarly approach should be the core concept for the development of degree programs in private security.

2. There is a need to assess the manpower, training, and educational requirements (managerial as well as technical level), both present and future, for the purpose of planning and developing academic programs.
3. There is a body of knowledge about the private security field sufficient to support realistic and meaningful 2-year, 4-year, and graduate-level college and university programs.

The following commentary is divided into two sub-areas—(1) associate degree programs, and (2) baccalaureate and graduate programs—to correspond with the differentiation made by most educational institutions.

Associate Degree Programs

A useful starting point in program planning is the Suggested Curriculum for Associate Programs contained in *Academic Guidelines for Security and Loss Prevention Programs in Community and Junior Colleges* (figure 15.1). A number of educators have indicated that this curriculum could serve as an excellent guide. Detailed course descriptions and other relevant information about designing and implementing programs can be found in the publication.

It would be inappropriate to recommend a set curriculum, because any program of private security education should be developed to meet the needs of local industry. Also, before developing and implementing degree programs, a review should be conducted of the assistance that colleges and universities should provide for training suggested (Chapter 2) and for seminars and courses (Standard 8.3). Immediate local industry needs can be better met by these forms of education. In any event, educational programs in appropriate forms should be designed with the specific needs of local industry in mind.

When developing degree programs, it may be difficult to identify the target population and to determine appropriate course content. However, it is strongly suggested that certificate, associate of arts, and associate of science degree programs be developed that include such subject matter as the following:

- Conducting security surveys.
- Historical, philosophical, and legal bases of the security field.
- Information security.
- Interviewing and report writing.
- Loss prevention techniques.
- Personnel security.
- Physical security.
- Principles and practices of fire prevention and safety.
- Supervision and leadership.

Figure 15.1
Suggested Curriculum for Associate Programs.

FIRST YEAR			
First Semester	Credits	Second Semester	Credits
English I	3	English II	3
General Psychology	3	Introduction to Sociology	3
Criminal and Civil Law I	3	Criminal and Civil Law II	3
Introduction to Security	3	Security Administration	3
Elective	3	Elective	3
	15		15
Electives:		Electives:	
Accounting I	3	Accounting II	3
Economics I	3	Economics II	3
Science I	3	Science II	3
Administration of Justice	3	Civil Rights & Civil Liberties	3
Principles of Interviewing	3	Report Writing	3
Industrial Relations	3		
SECOND YEAR			
First Semester	Credits	Second Semester	Credits
Fundamentals of Speech	3	Criminal Investigation	3
Social Problems	3	Criminology	3
Human Relations	3	Labor & Management Relations	3
Principles of Loss Prevention	3	Current Security Problems	3
Elective	3	Elective	3
	15		15
Electives:		Electives:	
Document & Personnel Security	3	Commercial/Retail Security	3
Business Mathematics	3	Field Practicum	3
Emergency Preparedness	3	Industrial Fire Protection	3
Environmental Security	3	Security Education	3
Physical Security	3	Special Security Problems	3
Safety & Fire Prevention	3		

- Unique security problems of hotels/motels, banks, manufacturing facilities, and so forth.

Baccalaureate and Graduate Programs

The lack of viable baccalaureate and graduate degree programs is both a handicap and an advantage. On the negative side, no curriculum model is presently available comparable to that which exists for associate degrees; therefore, each institution would have to develop its own curriculum without an historical frame of reference. This handicap, with proper research and planning, can turn into an advantage because no precedents exist that might need to be removed or modified during the developmental process. The following "Task Force Viewpoints for Development of Baccalaureate and

Graduate Programs in Private Security" (not listed in order of importance) are offered for consideration by educators:

Planning Phase

1. Each academic department of law enforcement/criminal justice should determine the number of graduates who are employed in the private security industry.

2. The academic departments of law enforcement/criminal justice should be the catalyst for development of security administration degree programs, but colleges of business need to be consulted and their courses incorporated into any degree programs. The disciplines of sociology, psychology, and law also should be included in the degree program.

3. New courses should be designed to incorporate the body of knowledge about private security subjects, rather than an effort made to adapt existing law enforcement/criminal justice courses to meet private security needs. For example, one or two law courses should be developed to relate pertinent legal aspects of the private security industry, rather than requiring security administration students to take a law course designed to prepare students for public law enforcement careers.

4. An advisory board, consisting of private security personnel, should be appointed to assist colleges and universities during the planning phase. This board should also remain active after the program is initiated.

5. Boards of higher education in each state should closely monitor all degree proposals in security administration to preclude proliferation of degree programs and to coordinate transfer arrangements between educational institutions.

6. Baccalaureate and master degree programs should be designed to prepare students for entrance-level or middle-management positions and not merely duplicate course offerings in certificate and associate degree programs. Some overlap and duplication may be necessary, but it should be kept to a minimum.

Implementation Phase

Each institution should determine the most appropriate way to implement private security curriculums, depending on available personnel and physical and financial resources; the following three-step process is recommended:

1. Introduce private security courses; then, if needed,
2. Develop private security minor; then, if needed,
3. Develop baccalaureate and/or master degree program(s).

Figure 15.2
Higher Educaton in Private Security.

Criteria	1977 PSTF	1984 Fischer Study	Journal of Security Administration
Some Private Security Courses (at least one)	77	N/R	N/R
Certificate Program	6	N/R	57
Associate Degree	22	N/R	84
Baccalaureate Degree	5	23	32
Master's Degree	0	8	14
Totals	110	31	187

N/R — not reported

CAUTION — all three studies were not conducted with the same criteria or methodology. For example, some studies include information on minors, concentrations, emphasis, options, and specialist while other studies excluded or did not report similar data.

Note — See Appendix C for additional detailed information about colleges/universities offering private security education.

Sources

1. PSTF — *Report of the Task Force on Private Security.* Washington, DC, U.S. Government Printing Office, 1977.
2. Fischer Study — Fischer, Robert J. "To What Degree?," *Security Management*, Vol. 28, No. 4, 1984.
3. Journal of Security Administration — "College Security Programs List (Revised)," *Journal of Security Administration*, Vol. 8, No. 1, 1985.

Bachelor of Science

It is recognized that many suggestions regarding curriculum design are arbitrary; however, it is established that a security degree should be interdisciplinary. The contemporary security professional must have a broad and balanced educational background consisting of general education, human relations, business practices, legal issues, loss prevention and control, and security. How else can the security professional function and be successful in accomplishing the tasks of preventing and controlling losses within an organizational environment? There must be an acute awareness and functional understanding of the role of security, how it "fits" particular and often unique organizational environments, and how it should be perceived and implemented within the context of various activities, e.g., marketing, production, personnel, maintenance, and/or retail, health services, transportation, manufacturing, etc.

The Bachelor of Science program in Security and Loss Prevention at Eastern Kentucky University is provided as an example of an interdisciplinary degree program.

MAJOR (Bachelor of Science)

Major Requirements .... 27 semester hours
SLP 110, 210, 225, 320, 333, 375, 385, 435, 465.

Supporting Course Requirements .... 39 semester hours
ACC 201, GBU 204, INS 378, PSY 202, FSE 221, 301[5], 305, 322, 410 and 412, CIS 212, CHE 101[4]

General Education Requirements .... 55 semester hours

Free Electives .... 7 semester hours

Total Curriculum Requirements .... 128 semester hours

Major Requirements (SLP)
Course Descriptions

SLP — SECURITY AND LOSS PREVENTION

110 Introduction to Security. (3)

The role of security, its application, and the security individual in modern society including an overview of the administration, personnel and physical aspects of the security field.

210 Security Technology and Hardware. (3)

An in-depth analysis and hands-on application of security hardware and technology; locks, security storage containers, electronic alarm devices, and alarm systems.

225 Legal Aspects of Fire and Loss Prevention Services. (3)

A study of legislative and legal decisions relating to personnel practices, employee safety and public protection in Fire, Safety and Loss Prevention services. Emphasizes the legal responsibilities, liabilities, and authority of the practitioner.

320 Emergency and Loss Prevention Services Management. (3)

Prerequisite: SLP 110 or instructor approval. An overview of organizational, administrative, and management practices in loss prevention and emergency services. Emphasis on supervision and leadership styles, motivation morale and organizational behavior.

325 White Collar Crime. (3)

An examination of white collar crime in America including its impact and modus operandi.

333 Comparative Security Programs. (3)

Prerequisite: SLP 225 or instructor approval. Study of security problems and practices in specific areas. Topical subjects would be in bank, campus, hospital and transportation security programs, etc.

385 Internal and External Security Controls. (3)

Evaluation of the major types and causes of internal and external losses that occur in business enterprises. Examination of techniques, motivations, and methods used by criminals within and without the business environment.

349 Cooperative Study. (1-8)

Prerequisite: department approval. Work, under faculty and field supervisors, in placements related to academic studies. One to eight hours credit per semester. Total hours: eight associate, sixteen baccalaureate. Minimum 80 hours work required for each academic credit.

375 Terrorism/Counterterrorism. (3)

A study of domestic, foreign and transnational terrorism, with emphasis on the philosophical bases, organization, equipment, and operations of terrorists groups. Role of law enforcement agencies in implementation of anti-terrorist measures is examined.

435 Topical Security Problems. (3)

Study of specific and current problems within society. Topical subjects include computer security, information security, organized crime, terroristic activities, etc.

445 Field Experience. (3-12)

Prerequisite: departmental approval. Field training is designed to broaden the educational experience through appropriate observational work assignments in cooperating agencies. May be retaken to a maximum of 12 hours.

455 Independent Study. (1-3)

Prerequisite: departmental approval. Individual reading and research on a problem or area within the field of security after student consultation with the instructor. Student must have the independent study proposal form approved by faculty supervisor and department chair prior to enrollment. May be retaken to a maximum of six hours.

465 Quantitative Loss Prevention Analysis. (3)

Concepts and procedures for quantitative loss prevention management techniques. Interpretation and application of loss prevention data and information for policy development and decision making.

Supporting Course Requirements

ACC 201 Principles of Accounting. (3)

Fundamental accounting relationships; completion of the accounting cycle; accounting process for merchandising enterprises; receivables, payables and inventories; deferrals, accruals, and intangible assets.

GBU 204 Legal Environment of Business. (3)

Law and the legal system; social forces that make the law; business responses to the social and legal environment. Focus on governmental regulations and federal regulatory agencies which impact business decision-making.

INS 378 Business Risk Management. (3)

Risk management as used by the business firm; basic functions of risk management; risk management decision-making as a corporate buyer of insurance.

PSY 202 Psychology as a Social Science. (3)

Introduction to concepts of psychology related to the social sciences, such as human development, learning, individual differences, personality development, adjustment, abnormal behavior, psychotherapy, social psychology, and applied psychology.

CIS 212 Indroduction to Data Processing Systems. (3)

Computer systems, including machine functions and computer organization; the symbolics of procedural analysis, data representation, computer mathematics, flowcharting, and computer programming with BASIC language, uses of computers in dynamic environments.

FSE 221 Fire Control I. (3)

Introduction to fire protection systems and their relationship to control and extinguishment. Study of extinguishing agents and their application. Concentration on fixed and portable carbon dioxide, dry chemical, dry powder, foam, and halogenated systems.

FSE 301 Emergency Medical Treatment. (5)

Effective emergency medical care in a variety of traumatic and medical emergencies. Content was developed by the Committee on Injuries of the American Medical Association. Nationally recognized for certification of emergency medical technicians.

FSE 305 Hazardous Materials. (3)

Study of hazardous materials in transportation, storage, and use. Chemical properties of hazardous materials relating to specific reaction, engineering controls, pre-emergency planning, combating, coordinating, and controlling a hazardous materials incident.

FSE 322 Fire Control II. (3)

Study of sprinkler systems, automatic fire detection systems, and municipal fire alarm systems.

FSE 410 Fire Prevention and Occupational Safety. (3)

Implementation of loss prevention techniques and programs. Fire and safety regulations; compliance with building codes and ordinances; insurance; relationship of occupational safety to fire prevention; economics of employee and property conservation.

412 Occupational and System Safety Management (3) A.

Prerequisite: FSE 410 or instructor approval. Principles and concepts of safety management and system safety. The relationship of safety law, employee management roles, ergonomics, psychological factors in safety motivation, safety reporting systems, risk analysis and identification and research of safety problems.

CHE 101 General Chemistry. (4)

Chemical bonding, structure of matter, chemical equilibrium and descriptive inorganic chemistry.

Program curriculum must be designed to meet the needs of the marketplace, i.e., graduates must possess the prerequisites required for employment in the field of security. Additionally, the curriculum must be dynamic, in that, it is constantly evaluated and changed to include technological innovations, emerging problems, and critical issues confronting the contemporary security practitioner.

Seminars, Workshops and Noncredit Courses

Practitioners, many who entered private security before private security education was generally available, cannot avail themselves of traditional educational opportunities. Thus, one of the best opportunities for colleges and universities to assist private security is through seminars, workshops, and noncredit courses.

Also, the rapidly changing technology for private security offers an opportunity for colleges and universities to serve as a catalyst to bring information about this technology to practitioners through specific training programs.

Participants at the First National Conference on Private Security (held at the University of Maryland in December 1975) resolved that "shared or cooperative training programs utilizing resources of private security, public law enforcement, education and training institutions (should) be pursued to meet the training needs of private security."

Training programs need to be developed for both operational and management personnel. Individual programs should be developed in cooperation with local private security employees, associations, and college and university officials. The following lists are provided for reference in developing seminars, workshops, and noncredit courses for private security.

Courses for Operational Personnel

- Arson investigation.
- Background investigation.
- Bomb threats.
- Civil disturbances.
- Cooperation with public law enforcement agencies.
- Employee security.
- Fire and safety.
- Firearms training.
- First aid.
- Internal theft investigations.
- Interviewing.
- Law and the private security industry.
- Note taking and report writing.
- Patrol methods.
- Physical security measures.
- Preliminary investigation.
- Public relations aspects of security services.
- Security training.
- Shoplifting prevention.
- Surveillance techniques.
- Visitor control.
- Terrorism.
- Computer/Information Security.

Courses for Management Personnel

- Crime prevention through environmental design.
- Disaster and emergency planning.
- Establishing a bomb threat plan.
- Establishing a more effective employee security awareness.
- Law/management relationship in security services.

- Private security/law enforcement relationship.
- Protection of key personnel.
- Security administration and management.
- Security problems discussion seminar.
- Strengthening business security.
- The role of the security administrator in research and development.

For this list, or any list, to be meaningful for course development, survey respondents should be requested to set priorities. If need in particular areas is established, several separate courses could become subjects contained in a longer seminar. For example, first aid, firearms training, and patrol methods might be covered in one seminar or workshop for operational personnel.

These short courses can fill the present short-range need for educational opportunities while the academic programs recommended in other standards are being developed. Later, these courses can support the regular academic classes and, at the same time, give educational institutions a mechanism to provide information to meet continually changing private security situations.

The opportunities for innovative use of seminars and short courses to meet private security needs are apparent. Educational institutions, in cooperation with the private security industry, should take the initiative to provide the physical and personnel resources to implement this standard.

Summary

In summary, higher education and training provide the foundation and impetus for improving the professional competence and image of the security industry. There should not be any question as to the value and importance of education and professional training for the security practitioner. The complexities and dynamics of society, the workplace, and the world dictate the need for security professionals who are university educated and prepared to confront a myriad of changing issues and problems.

Selected References

1. Calder, James D. (chairman, resolutions committee). "Resolutions of First National Conference on Private Security" (final draft). College Park: University of Maryland, December 1975.
2. Conrad, John J. (editor). "Department of Law Enforcement, Western Illinois University, Newsletter #1, Academic Year 1975–1976."

3. Cunningham, William C. and Taylor, Todd H. *Private Security and Police in America*, Portland, OR: Chancellor Press, 1985.
4. Fauth, Kenneth G. "The Need for Security Education at the Post-secondary Level," unpublished doctoral dissertation. Bloomington: University of Indiana, 1975.
5. Fischer, Robert James. "The Development of Baccalaureate Degree Programs in Private Security 1957–1980," unpublished doctoral disseration. Carbondale, Illinois: Southern Illinois University, December, 1981.
6. ________. "College Security Program List (Revised)," *Journal of Security Administration*, Vol. 8, No. 1, 1985.
7. ________. "To What Degree?," *Security Management*, Vol. 28, No. 4, 1984.
8. Friend, Bernard D. "Profile of the Physical Security Officer: Specialization or Professionalism," unpublished master's thesis. East Lansing: Michigan State University, 1968.
9. International Association of Chiefs of Police. *Law Enforcement and Criminal Justice Education Directory 1975–1976*. Gaithersburg, Md.: International Association of Chiefs of Police, 1975.
10. Kingsbury, Arthur A. (project director). *Academic Guidelines for Security and Loss Prevention Programs in Community and Junior Colleges*. Washington, D.C.: American Society for Industrial Security/ASIS Foundation, Inc., 1972.
11. ________. "Macomb College Looks at Security Education," *Industrial Security*, August 1970.
12. ________. "Security Education," *Security Management*. September 1974.
13. Larkins, Hayes Carlton. "A Survey of Experiences, Activities, and Views of the Industrial Security Administration Graduates of Michigan State University," unpublished master's thesis. East Lansing: Michigan State University, 1966.
14. Moore, Merlyn Douglas. "A Study of the Placement and Utilization Patterns and Views of the Criminal Justice Graduates of Michigan State University," unpublished doctoral dissertation. East Lansing: Michigan State University, 1972.
15. Van Meter, Clifford W. "Perceptions of Selected Criminal Justice Graduates, Faculty, and Police Chiefs on the Impact of Education on Job Performance, Promotion and Job Satisfaction," unpublished doctoral disseration. Carbondale, Illinois: Southern Illinois University, 1982.

Discussion Questions

1. List and discuss the three resolutions passed at the First National Conference on Private Security regarding private security education.
2. How should a college or university go about the development of a security curriculum?
3. Discuss the attributes of an interdisciplinary approach to security education.

Chapter 16
Legal Aspects of Private Security

Our legal system attempts to strike a balance between the rights of persons and private organizations to protect themselves and their property and the rights of private citizens to be free from the power of others. The attempt to balance conflicting interests is nowhere more apparent than in the field of private security.

On the one hand, the private sector uses security employees to protect their lives, property and customers from the mugger, shoplifter, pickpocket, hijacker, embezzler, arsonist, vandal and other threats. On the other hand, all citizens are entitled to be free from assault and battery by others, unlawful detention or arrest, injury to reputation, intrusion into personal privacy and illegal invasions of one's land, dwelling or personal property.

In order to perform effectively, private security personnel must, in many instances, walk a tight-rope between permissible protective activities and unlawful interferences with the rights of private citizens. The precise limits of the authority of private security personnel are not clearly spelled out in any one set of legal materials. Rather, one must look at a number of sources in order to define, even in a rough way, the dividing line between proper and improper private security behavior. These sources are briefly discussed below.

Constitutional Law

The Federal Constitution places many limitations on the conduct of government officials, including police and quasi-police agencies and other components of the total criminal justice system. In general, however, the Constitution says little about the rights of private citizens in their relationship to other citizens. Most Constitutional rights of an individual relate to governmental or state action and not to activities of other private persons or corporations.

Constitutional limitations can, however, apply to the conduct of private security personnel when private security personnel act in concert with public law enforcement officials or as agents to obtain evidence with the intent of furnishing it to such officials for use in a prosecution.

Basically, a security officer may possess one of the three kinds of power and authority. An authority identical to that possessed by a citizen or a property owner is by far the most common power held by security officers. Less common is an authority that is obtained by deputization or commissioning from a public law enforcement agency. The third type of authority possessed by some security officers is a mixture of the powers of a civilian with certain special prerogatives added by statute, ordinance or governmental regulation. A full-time police officer working as a part-time security officer would be in this third area. Deputization is the most frequent legal action which vests the private citizen or private security official with full police powers (as well as the restrictions and limitations on such powers). The extent of permissible deputization is limited. In Maryland, for example, the governor is able to appoint "special policemen" to work for private businesses with full police power limited to the property of the requesting business. (Md.Ann.Code art. 41, § 60). Oregon has a similar code but only in the railroad and steamboat industries. (O.R.S. 148, 210). Deputization does not always give full police powers, but usually makes the deputized individual subject to constitutional restrictions.[1]

As constitutional restrictions usually apply to state actions only, such restrictions do not generally pertain to private security activities. Only in rare instances will private police action be classified as state action. The question of whether or not licensing private security personnel constitutes state action is raised in *Weyadt* v. *Mason's Store Inc.*, 279 F.Supp. 283, W.D.Pa. (1968). The argument was the same as that in *Burton* v. *Wilmington Parking Authority*, 365 U.S. 715 (1961), where the Supreme Court held that there were sufficient grounds for a restaurant owner to be liable for racial discrimination, because the state granted a lease to the restaurant. That reasoning was not followed in *Weyadt* v. *Mason's*. It was held that although a private detective of a store was licensed under the Pennsylvania Private Detective Act and was acting under "color of the law," the law is not a deputization law and does not invest the licensee with the authority of state law. Thus, the private detective of the store was only acting with the authority of a "private citizen." When private security personnel are hired on a contractual basis by a public authority, they are in fact acting with authority of state law and do have the imposition of constitutional restrictions upon the exercise of power (*Williams* v. *United States* 341 U.S. 97 [1951]). This applies to the area of arrest as well as other categories of legal involvement of private security officials.

Criminal Law

Criminal codes offer both a source of power and restraint for the private security officer. In criminal law, an action is defined as a "social harm" for which the offender is answerable to society (not to an individual as in torts) and is punishable by law.[2] Criminal law operates as a deterrent to the extent that the law is known, that the consequences of being convicted are sufficient, and that the criminal justice system operates effectively in imposing sanctions. Because intent to commit a crime is required, crimes are narrowly defined and the prosecution must prove guilt beyond a reasonable doubt. The criminal law can be best seen as establishing outer limits on behavior rather than as a day-to-day regulatory device.

Two concepts are important if private security personnel are to acquire a working knowledge of criminal law. First is the legal maxim that everyone is presumed to know the laws of the state and nation. Of course, this legal premise is not valid when tested on a non-legal, practical basis. Even lawyers are not able to recite all of the laws of a state. Nevertheless, this rule prevails and is applied in courts of law.

The second legal concept is that a law must be so clear and understandable that an ordinary person will know what conduct is prohibited.

Before a person can be convicted of violating one of the specific laws of the state, the prosecution must establish beyond a reasonable doubt each of the elements of the offense. For example, when prosecuting a person charged for violation of the common law offense of rape, the state must prove these elements: a) unlawful, b) carnal knowledge, c) female, d) by force, e) against her will. If one element is not proved even though the others are shown beyond a reasonable doubt, there can be no conviction of that offense. There is a possibility, however, that the person can be convicted of a lesser included offense.

Tort Law

The primary source for the authority of private security officers and the limitations on such authority is tort law. The law of torts is found in both legislation and court-developed common law. There is no one body of tort law; it varies from state to state. However, there is an ongoing attempt to achieve some conformity through various model laws and the *Restatement of Torts*, which is published by the American Law Institute. Tort law is defined as a body of law that governs the civil relationships between people.[3] It defines and creates causes of actions permitting one person to remedy the wrongs committed against him by another, and has the effect of restraining conduct by making the wrongdoer aware of the one injured. These remedies may either be equitable (the enjoining of certain conduct) or legal (the recovery of money damages for injuries received). Tort law differs from criminal law in that private parties are suing, and the suing party is seeking relief for himself and not punishment of the offending party.

To a certain extent, tort law defines privileges and immunities that offer a source of authority for private conduct. Early rules of arrest, prevention of crime, self-defense, defense of others and of property have their basis in common law tort principles. Further, tort law protects an individual's person and property from injurious conduct, his reputation from disparagement, his privacy from unreasonable exposure, and his mental well-being from emotional distress. Conduct which harms another and violates norms of reasonableness is generally actionable if done without privilege or immunity.

Tort law does not provide specific authority for private security officers, but it does define at least some limits on the conduct of private security personnel. It allows for an injured party to bring a lawsuit for damages and injuries caused by "tortious" conduct by private security officers. The courts follow precedents when established, and create remedies to fit novel cases. Typically these remedies come from common law. Thus, tort law restrains the authority of private security officers only by the threat of a subsequent lawsuit, and provides general parameters on reasonable conduct through case law precedents.

Tort law provides that a civil action for damages may be maintained by one party against another whenever the former has suffered a compensable injury or loss by reason of the act of the latter. Therefore, an act or omission which causes such injuries or losses to another is known as a civil wrong or a tort. Generally speaking, in order for the injured party to recover damages from the tortfeasor such injured party must prove the tortfeasor acted in a negligent manner contrary to his duty to act prudently, and that the negligent act was the cause of certain compensable losses or injuries incurred by the victim.

The purpose of a civil action is to provide reasonable compensation to an injured party for the damages incurred by him due to the acts of another. Therefore, as opposed to criminal prosecutions where the intent of the accused, *mens rea*, is a major element of the offense in civil actions, there is much less concern for the mental state of the defendant tortfeasor. The major concern is that he be required to pay for the damages he caused another. Thus in tort cases, there is usually no element of *mens rea* so the victim need not prove that the tortfeasor intentionally damaged the victim.

Most civil actions, then, are based not upon any claim that the defendant intended harm, but that the defendant was negligent in his conduct and it was this negligence which occasioned the loss for which a recovery is sought. Negligence or the absence of due diligence is the key to most cases involving a security officer. The law requires that all persons conduct themselves with due regard for the safety and rights of others, and the failure to do so constitutes negligence. The standard by which particular conduct is tested to determine whether such conduct constitutes negligence is that of the "reasonable man." That is, would a reasonable person of ordinary prudence have acted similarly under the same or similar circumstances? Of course, this standard is higher when the conduct of certain professionals is in question. One who holds himself out to the public as being an expert in a particular field thereby assumes the duty to conduct his activities in accordance with the generally accepted community standards of others in the same profession. Therefore, one who holds himself out to the public as having special knowledge or expertise in the field of private security, for instance, will have his conduct measured against this higher standard rather than by the standard of the "common man" who professes no such expertise.

Contract Law

There are several types of contractual arrangements which are important to the scope of authority of private security personnel and the component areas of service which are offered by a full service security agency. The performance of the security officer has been the dominant concern of contract law regarding security services since World War II. In addition, improvement in electronic alarm systems and seemingly reduced costs have led many busineses to install them in the last decade. This movement toward more sophisticated electronic alarms and their extravagant advertising claims have focused attention in contract law to the various alarm systems and their relationships to a business enterprise.

The terms of a contract between a business enterprise and a security service may limit the private security officer's authority and define more stringent standards of behavior than are defined in other bodies of law. The contract between the security agent and the hiring company will usually define the respective liabilities of all parties for the business enterprise utilizing the contractual security service. If there is harm to a third party, the contract will usually establish who is to be responsible and who is to carry insurance for which risks. However, the courts have, on occasion in suits by third parties, held a person liable even though a contract said that another was to be responsible. (See *Annotation*, "Liabilities of One Contracting for Private Police or Security Services for Acts of Personnel Supplied," 38 ALR 3rd 1332 [1971].) In addition, union contracts may impose restraints on employers (and thus on private security personnel) in such areas as search of employee lockers and belongings and the conduct of investigations into employee wrongdoing.

Statutory Provisions

Restraints on the conduct of private security personnel may also be found in a variety of state and local statutes, rules, ordinances and regulations. Much of this legislation is in the form of licensing or registration statutes which place requirements on qualifications of security personnel to obtain or retain a license or permit. To some extent, these statutes also delineate proper forms of conduct and restrain other types of conduct.

Many of these regulations provide for suspension or revocation of a license, and include provisions requiring surety bonds or insurance from the security agency to protect clients and employees. Special powers may be granted to private security personnel, such as the right to carry a weapon. The statutory provisions vary by state and locality, and enforcement procedures vary even more.

Special attention needs to be given to four problem areas in private security: (1) use of force, (2) interrogation and questioning, (3) search, and (4) a comparison between public and private police.

1) *Use of force.* Any activity which is performed by a security officer which interferes with the rights of other persons, unless there is privilege or consent, places the security officer in a position of possibly being held liable. The primary legal basis for many activities performed by a security officer is consent. If a security officer requests a person to do a certain act, i.e., leave an amusement park and the customer does so voluntarily, the customer has consented to leave the park. Consent cannot be coerced, however, and it may also be limited in its scope. When consent is not given, there are various privileges that usually provide the legal basis for enforcement activities.

   Two privileges which usually are important are the right of a real property owner to prevent trespasses on his property and the privilege of self-defense. A person may use reasonable force against someone who appears to be about to inflict physical harm. Generally, however, only an amount of force reasonably necessary to realize the legitimate purposes of the privilege is allowed. If excessive or unreasonable force is used, not only is one liable for the torts resulting from this excessive force (usually battery), but the original privilege is also lost.

   There are no clear rules as to what force is allowable in given situations. Usually "reasonableness" controls, and what is reasonable depends on the nature of the interest being protected, the nature of the act being resisted, and the particular facts in a given situation. To add to the confusion, the amount of force allowed is different depending upon which privilege is being invoked. Where property rights are involved, a request for voluntary cooperation (consent) should precede the use of any force. Similarly, when only property is at stake, the use of deadly force (e.g., a gun) is impermissible unless the threat to property also threatens life. The law places a higher value upon human safety and life than upon mere rights in property and it is the accepted rule in law that there is no privilege to use any force calculated to cause death or serious bodily injury to repel the threat to land or chattels, unless there is also such a threat to the property owner's or his representative's personal safety as to justify self-defense. This rule would likewise apply to the privileges granted in all shoplifting cases.

2) *Interrogation and questioning.* As long as a person is legally detained, there is no absolute ban on simply asking questions. Courts have held that an improper interrogation is not itself a tort, whereas an illegal arrest or an illegal search or seizure may be. (295 F.Supp., 1184, 1186 [D. Md., 1969].) However, there are still certain limits imposed by state law on the methods of interrogation. The suspect is under no legal obligation to answer the questions and has the right to remain silent. The private security officer who is *not* operating under special sanctions and commissions of a state is not obligated to advise the suspect of these rights.

   Some indirect control over interrogation methods is exerted by virtue of contract law: any releases, promises, or agreements signed or entered into as a result of coercion or duress would be unenforceable. Indirect control also exists by virtue of decisions by some courts that confessions or admissions obtained by coercion, force, and (sometimes) promises, are inadmissible in subsequent criminal proceedings against the suspect.

   The use of physical force or threats of physical force to coerce answers is prohibited. Such threats or force would be tortious, either directly as assault or battery, or indirectly as constituting unreasonable exercise of the detention or arrest privilege involved. The legality of any interrogation or questioning will turn on the manner in which the questioning is conducted, and the standard for the manner of questioning is basically "reasonableness."

3) *Search.* There is a vast difference between searches conducted by law enforcement officers and searches made by private persons. And unless a private security officer is acting in concert with police officials or is acting under the authority of state or local laws which make him a "quasi" police officer, the private security officer is just like any other private person in this regard.

   A police officer is most often concerned with whether the results of his search will be admissible as evidence in a criminal prosecution. The fourth amendment and its companion, the Exclusionary Rule, require that such governmental searches be based on probable cause. A police officer may also be, but very rarely is, held liable in civil damages to the victim of an unlawful search, either on the basis of invasion of privacy or under a civil rights statute.

The fourth amendment does not, however, apply to searches by private persons. Evidence discovered by means of a private search is ordinarily admissible in a criminal prosecution even though there was no basis at all for the search other than mere curiosity or snooping. A private security officer, though, must be more concerned with possible civil, and even criminal, liability which may result from an illegal or unreasonable search.

The legality of a search, like the legality of interrogation or questioning, is usually inseparable from the legality of the initial detention of the suspect. Following this logic, if there was no probable cause for the detention, then any search would also be illegal. But assuming a legitimate basis for detention, the question is whether a private security officer has the legal power to search a suspect's person, including any purse, briefcase, or other items the suspect is carrying.

Often, consent will render the search valid, particularly if the suspect physically cooperates in the search at the request of the arresting private security officer. Without consent, some legal privilege or right must be found to justify a search. However, the privileges for private police which might allow searches are not clear-cut, nor is any single privilege, like self-defense, expansive enough to sanction a wide range of searches in a number of different situations. The law of searches in the private sector has simply not been developed as it has in the public police sector.

The common-law right of self-defense might justify reasonable searches for weapons, but only where there is reasonable ground to fear imminent attack by use of a concealed weapon. Under the common law, the arresting individual was empowered to search a suspect who is already under arrest if the arresting individual "has reason to believe that he has on or about him any offensive weapons or incriminating articles." (See Perkins, *supra*, note 304 at 261, Warner, "The Uniform Arrest Act," 28 Va.L.Rev., 315, 324 (1942); *United States* v. *Viale*, 312 F.2d 595 (2d Cir., 1963); Restatement of Torts (2d) 132, comment d.) However, this power was limited to cases of formal arrest (i.e., where the person will be turned over to the authorities), not mere detention. A similar right is provided by state statutes specially authorizing private citizens when making an arrest to seize weapons from the arrested person. The common law privilege of recapturing chattels wrongfully taken seems to support a search about the person for such goods or chattels in a non-arrest situation; however, in states which allow this practice, its use is quite limited. Most merchant detention privileges would not support such a search and some states even prohibit it by statute.

4) *Comparison between public and private police.* Under state tort laws and state statutes, a public police officer has significantly more powers than a non-deputized private police officer. A public police officer can obtain and serve a search or arrest warrant. Even without a warrant, a public police officer may arrest or detain a suspect in all of the situations in which a private citizen could, plus many more. For example, a public police officer can usually arrest for the commission of a felony as long as he has reasonable grounds for believing a felony has been committed. In contrast, a private citizen may usually arrest for a felony only when, in fact, the felony has been committed. Further, not only is a public police officer usually vested with the same powers as a merchant or his agent to detain shoplifters, he is often empowered to stop or temporarily detain persons suspected of other crimes. A public police officer is usually granted specific statutory power to conduct, incident to an arrest, a search for weapons and contraband, and to frisk any temporarily detained person for a weapon if probable cause exists.

Further, a public police officer would seem to have at least the same powers as a private police officer to take actions short of arrest, for example, expelling intruders or persons causing disturbance from private property. He is as capable of acting on behalf of the owner as any other agent of the owner. Moreover, he is often given specific statutory authority to take such action in particular situations.

Moreover, as a practical matter, enforcement of restrictions on his activities by means of civil or criminal lawsuit is much less likely, and a police officer is likely to encounter much less resistance to requests for voluntary cooperation. Finally, in some states it is illegal to resist an arrest by an officer of the law, even if the arrest is illegal.

On the other hand, the public police officer is subject to various restrictions imposed by the Federal Constitution which, so far, have not been generally imposed upon private detectives and guards. In those areas controlled by the Constitution, a private person may be less restricted and thus have more power than the public police.

First, as discussed earlier, a public police officer's power to arrest without a warrant is restricted by the Fourth Amendment to situations in which he has probable cause, regardless of whether the crime was in fact committed. In contrast, lack of probable cause in a citizen's arrest would be excused if the felony had in fact been committed.

Second, the Supreme Court has held (in *Chimel* v. *California*) that warrantless searches incident to a valid arrest may extend only into the area within which an arrestee might reach a weapon or destructible evidence. And while the court in *Terry* v. *Ohio* sanctioned the warrantless "stop and frisk" of a person engaged in suspicious activities, in *Sibron* v. *New York* the court strictly limited the searches incident thereto to a pat-down for suspected weapons—not for evidence. These constitutional limits upon the scope of searches incident to arrests and detentions have not yet been imposed on private searches, and are generally more restrictive than those the courts have so far applied by tort law to private searches. On the other hand, tort law governing private searches of the person is neither clear nor well developed. Thus, since both the Supreme Court's recent Fourth Amendment decisions and the tort law rely heavily upon "reasonableness," the two areas of law may well coincide—without directly imposing "constitutional" standards upon the activities of private security officers. Indeed, tort law might eventually impose more restrictive standards on private police than are now applied by constitutional law to public police.

Third, the Supreme Court has placed restrictions upon police methods of interrogating suspects to ensure that any confessions are voluntary and not "coerced." The court has enforced these standards by excluding evidence obtained by improper interrogations from subsequent criminal prosecutions of the suspect. Many courts have rendered these constitutional standards applicable indirectly to private persons by excluding any "coerced" confessions from criminal prosecutions, regardless of whether the source of coercion was public or private.

However, in *Miranda* v. *Arizona*, the Supreme Court imposed an additional requirement for the admissibility of incriminating statements. Before interrogation, the suspect must be informed of his right to remain silent and to obtain the assistance of counsel. This requirement clearly goes beyond what is required of private security personnel by virtue of tort law or by the admissibility standards applied in such decisions.

In summary, the public police may have greater arrest, search, and interrogation powers under state law, but they are also subject to constitutional restrictions which, so far, have not been imposed upon private police. Nevertheless, these constitutional restrictions may not, on occasion, be significantly different from the restrictions imposed on private police by tort law.

Current Legal Trends

According to a recent report in *Security World*, payments awarded in security related lawsuits have been increasing at the rate of 300 percent per year since 1967.[4] These lawsuits have affected a wide range of businesses and organizations ranging from retailers, health care facilities, financial institutions, schools, hotels/motels, civic groups, and municipal governments. It appears that the majority of cases revolve around the question(s) of inadequate security, improper security practices, and failure to prevent crimes. The litigants bringing suit for negligent actions include victims of criminal actions that occurred in parking lots, motel rooms, hallways, bars, restaurants, apartments, etc.; employees claiming they were subjected to improper security actions by their employer; customers charging that they were subjected to improper security procedures ranging from searches and seizures, slander, harassment, etc.; and lawsuits by customers, invitees, and others over failure to perform adequate employment practices such as training, pre-employment selection procedures, and employee controls.

Without question, organizations of whatever type are being forced to consider and recognize that certain acts and/or omissions in regard to adequate security controls can result in a major lawsuit. Any step taken to loosen or tighten security may be a crucial factor in the determination of liability by a court of law. The following cases are cited to provide a brief overview of lawsuits in selected areas:

1. *Employee Violence*

   Ruiz v. Heldt Brothers Trucks, 229th District Court of Texas, Starr County, 1985, held a Texas employer liable for $4 million in punitive damages for the killing of one employee by another on a work site.

   Duarte v. Bayless, Inc., Superior Court, Maricopa County, Arizona, 1985, awarded $500,000.00 to the survivors when a security guard shot and killed a suspect in a shoplifting case.

2. *Negligent Security*

   Martin v. Norm's Restaurants, Inc., Los Angeles Superior Court, 1985, awarded $733,339.00 to survivors in a negligent security suit after Martin was shot and killed. It was determined that security was inadequate.

   Hughes v. Jardel Co., Superior Court, New Castle County, Delaware, 1985, awarded $530,000.00 in compensatory and $250,000.00 in punitive damages to a raped store employee after the security firm responsible for security testified it had recommended more security.

3. *Employee Drug Testing*

   McLeod v. City of Detroit, U.S. District Court, Eastern District of Michigan, Southern Division, August, 1985, upheld Detroit's program of urinalysis of firefighter candidates for marijuana, and rejection on the basis of positive results.

   Turner v. Fraternal Order of Police, District of Columbia Court of Appeals, November, 1985, upheld Washington, D.C. Police Department's order that police officers would have to undergo urinalysis on the basis of suspected drug use.

4. *Crimes by Employees*

Victoria v. Kaiser Foundation Hospitals, Supreme Court of California, December, 1985, held that a hospital patient raped by an orderly could sue the hospital for negligent hiring and supervision.

Henley v. Prince George's County, Court of Appeals of Maryland, February, 1986, ordered to trial a suit against a contractor and others over the murder of a 12-year-old boy by an employee who was hired to provide security.

Doe v. Durtschi, Supreme Court of Idaho, February, 1986, ordered a trial in a suit against a school district for negligent hiring of a school teacher. The teacher had previously molested fourth grade students.

5. *Private Security and Public Police*

Sovary v. Los Angeles Police Department, 86 D.A.R. 361, California Court of Appeals, Second Appellate District (1986) held that the Los Angeles police may be liable for an injury in a shopping mall. The police advised members of a shopping plaza against hiring private security, and promised instead to provide continuous bicycle and foot patrol.

In Re KUMA K-9 Security Inc., No. 148, Superior Court of Pennsylvania, March 19, 1986, held that a private detective agency's license should not be renewed while the agency retained a high-ranking public police officer as a consultant.

Dr. Lawrence W. Sherman, Director of the Security Law Institute and executive editor of the *Security Law Newsletter*, has long advocated the need for general standards in regard to determining the forseeability of crime and the adequacy of security on premises open to the public. In 1984, Dr. Sherman published a draft of standards drawn from discussions with security professionals and related professional meetings.

The standards, as listed below, suggest a guideline and process for (1) reducing the risk of crime against customers, invitees, vendors, etc., on premises open to the public, and (2) reducing the uncertainty that owners and operators of such premises face in the determination of liability under civil law for failing to prevent any crimes that may occur.[5]

STANDARDS

I. DUTY TO PREVENT CRIME

Operators of premises open to the public should recognize a duty to

A. take reasonable steps
B. to reduce the risk of
C. reasonably foreseeable types of crime being committed
D. on or immediately approaching the premises
E. by unknown third parties and
F. premises employees, against

G. all invitees.

The scope of this duty should not be limited as to the type of organization owning or operating public premises, nor by governmental immunity.

II. FORESEEABILITY OF CRIME

A. It is reasonable to foresee "normal" crimes committed
   1. against invitees' persons
   2. by strangers to them, if
   3. there have been *any* stranger crimes against persons
   4. within a one-mile radius of the premises in question, or
   5. minor crimes suggesting an order maintenance problem on the premises,
   6. within the last two years.

B. In order to determine whether crimes by strangers are reasonably foreseeable, premises operators should take affirmative steps to keep informed of local criminal activity, including
   1. regularly requesting the police department to advise the operator whether there have been any stranger-to-stranger crimes against persons within a one-mile radius of the premises.

C. It is not generally reasonable to expect a premises operator to foresee stranger crimes against invitees if local police cannot or do not provide local crime information.

D. It is reasonable to foresee crimes committed
   1. by a premises employee against invitees' persons, if the employee has
   2. committed any act of violence while on the premises.

E. It is generally not reasonable, nor is it wise and fair public policy, for premises operators to attempt to determine the risk of an employee committing a crime by attempting to gain access to the employee's prior criminal record, if any.

III. ADEQUACY OF SECURITY

A. It is reasonable to determine whether security to protect invitees was adequate at any specific premises only in light of all the particular facts concerning the prior history of crime at and near the location and of the security measures taken to prevent crime.
   1. It is not generally reasonable to determine the adequacy of security by applying a general rule about a specific security measure being required for all premises, or all premises of a certain kind.
   2. It is not generally reasonable to determine the adequacy of

security at a specific location in relation to the "prevailing practices" of similar premises in the area or nationally.

B . A reasonable standard of care in protecting invitees from crime requires that no reduction in the level of effort, nor change in major strategy, of security be undertaken without a documented, professional analysis of the changing local crime risk factors or advances in security methods that can reasonably justify changes in security.

C. A reasonable standard of care also requires that if a substantial increase occurs in the rate at which invitees are victimized by crime upon the premises, all security measures should be
   1. professionally reviewed, and
   2. probably increased in their level of effort.

D. The strategy and level of effort of security to protect invitees from crime on any premises should be
   1. reviewed and certified as adequate at regular intervals,
   2. by two or more security professionals,
   3. whose qualifications would be likely to be adequate for establishing the admissibility of their testimony as security experts in civil trials, and
   4. who have no financial interest in the sale of any of the security products or services which might be relevant to the premises.

E. All security professionals and operators of public premises should cooperate with scientific research efforts to determine more precisely the relative effectiveness of different kinds of security measures in preventing different kinds of crime on different kinds of premises.

Summary

While arrest or police powers are not generally conferred on private security personnel by state statute, in some states enabling legislation or county and local ordinances to grant special police powers to licensed private security personnel under specific conditions. In addition, forty-five states, through state statute or common law, permit arrests by private citizens, and the majority of states have enacted anti-shoplifting statutes which permit detention of suspected shoplifters by private security agents of a merchant. In all of these instances of special police powers, citizen arrest privileges and shoplifting detention statutes, there is great variation among states as to the privileges conferred and legal restraints imposed on the conduct of private security officers.

In recent years, there has been an explosion of civil court cases over inadequate and/or questionable security practices. Such cases have revealed a critical need for a compendium of standards which can serve as a prescriptive guideline of reasonable, consistent measures of security to be utilized by property owners, courts, and security personnel.

Discussion Questions

1. What are the types of power and authority that a security officer may possibly possess?
2. Briefly discuss the significance and relationship of tort law to the security officer.
3. Briefly outline and discuss the four problem areas of private security which are relevant to the practitioner.
4. Define *mens rea* and its relation to tort and criminal law.
5. Discuss the "standards of security" suggested by Dr. Lawrence W. Sherman.

Notes

1. W.J. Bird, J.S. Kakalik, S. Wildhorn, et al. *The Law and the Private Police*, The Rand Corporation (Santa Monica, California), February 1972, Vol. II, p. 13.
2. Rollin M. Perkins. *Criminal Law*. 2nd ed. Foundation Press (Mineola, New York), 1969, p. 23.
3. William L. Prosser and Young B. Smith, *TORTS*. 4th ed. Foundation Press (Mineola, New York), 1967, pp. 1-2.
4. Lawrence Sherman. "Liability and the Standards Issue," *Security World*. January 1985, p. 37.
5. Dr. Lawrence W. Sherman, *Protecting Customers From Crime*. Draft Standards with Commentary for Determining the Foreseeability of Crime and the Adequacy of Security on Premises Open to the Public, Security Law Institute. Washington, D.C., September 1984.

Chapter 17
Career Orientation and Conclusions

In a free enterprise system the best indicator of success is sustained growth. By this criterion, private security is a success. During the 1970's the best data available indicate that private security grew at an annual rate of 10% per year. Few segments of the American economy have, during that same period, been able to maintain that rate of growth.

Two broad hypotheses have been used to explain this growth. One is that the private security industry has grown because of the increase in crime in the United States and the inability of public law enforcement to stop this increase. The second is that the quality of private security services has improved and that this has been recognized by industry, which has resulted in greater utilization of private security services.

The truth probably lies somewhere in between. For example, an argument often used to explain data showing increases in the crime rate is that it is better reporting of crimes that accounts for much of the increase. On the other hand, some argue that the quality of private security has not really improved, but that it has simply become an increasingly necessary part of doing business.

Another position is that the real factor in the growth is government regulation. The security requirements for doing business with the government (primarily federal) have increased, and industry has responded by increasing security for the simple reason that without it they could not get government contracts.

Taking a "crystal ball" look at the future until the end of the twentieth century seems to be reasonable. If crime continues to increase and the quality of private security continues to improve, then the growth rate will continue. Although there appears to be no hard data on the subject, most private security executives speculate that the growth will not continue at the 10%-a-year rate. The best estimates are at about the 5% level. At first glance this might be discouraging, but in recent years the American economy has been growing at less than a 5%-per-year rate. Thus, planning a career in private security would seem to be an excellent idea.

Public vs. Private Security – Some Data

Throughout this text constant references have been made to the important contributions of private security in the overall effort toward crime prevention. A significant contribution of data is contained in a book titled *Private Security and Police in America* published in 1985. The authors, William C. Cunningham and Todd H. Taylor, conducted extensive research to make their projections.

Two major findings further supported research conducted by the Rand Corporation in the early 1970's and the Private Security Task Force in the middle 1970's. These related to the number of private security personnel and total expenditures of private security compared to public law enforcement.

The projections were based on careful analysis of data collected by the Bureau of Justice Statistics, the Bureau of Labor Statistics, and the Bureau of the Census.

Estimated private sector security employment in 1982 was 1.10 million, and Table 17.1 provides detailed information about the number of per-

Table 17-1
Hallcrest Estimated Private Sector* Security Employment in the U.S. 1982.
*Excludes Government (Civil and Military) Security Workers

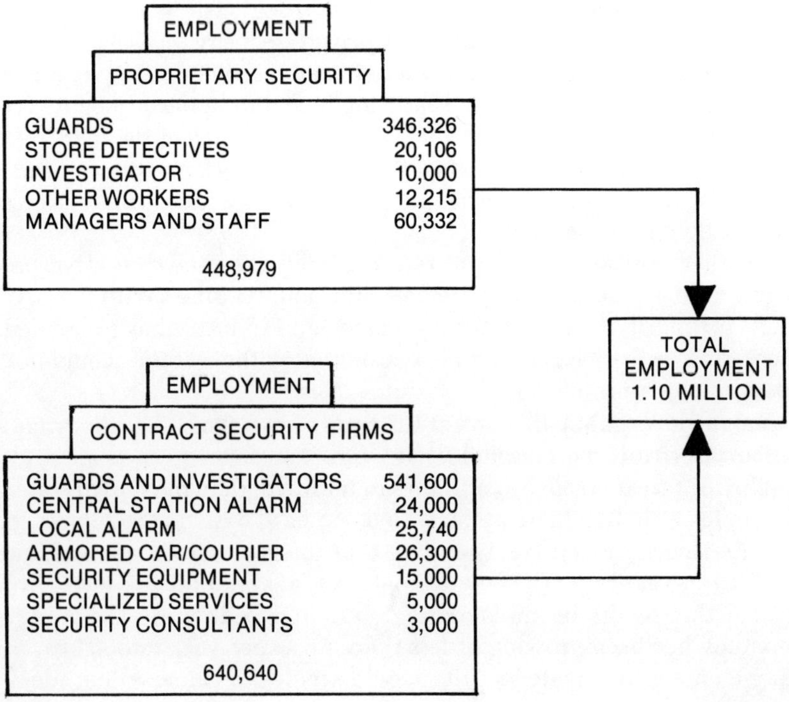

EMPLOYMENT — PROPRIETARY SECURITY	
GUARDS	346,326
STORE DETECTIVES	20,106
INVESTIGATOR	10,000
OTHER WORKERS	12,215
MANAGERS AND STAFF	60,332
	448,979

EMPLOYMENT — CONTRACT SECURITY FIRMS	
GUARDS AND INVESTIGATORS	541,600
CENTRAL STATION ALARM	24,000
LOCAL ALARM	25,740
ARMORED CAR/COURIER	26,300
SECURITY EQUIPMENT	15,000
SPECIALIZED SERVICES	5,000
SECURITY CONSULTANTS	3,000
	640,640

TOTAL EMPLOYMENT 1.10 MILLION

sons in proprietary and contract security. This table does not specifically include employment data for public police, but other research indicated that the total was under 600,000.

Detailed information about the estimated annual expenditures for protection services provides another perspective on the issue. For example, 1979 expenditures for police protection were estimated at $13.8 billion while estimates for private protection in 1980 were $21.7 billion. Table 17.2 provides additional detailed information.

Table 17-2
Summary of Estimated Annual Expenditures for Protected Services.

POLICE PROTECTION (1979)[1]	**$ BILLION**	%
Federal	$ 1.0	14.1%
State	2.1	14.4%
Local	9.8	71.5%
TOTAL	$13.8 billion	100%
PRIVATE PROTECTION (1980)[2, 3]		
Industrial/Manufacturing	$ 5.9	27.6%
Retailing	3.8	17.4%
Government Installations	3.3	15.9%
Financial Institutions	1.9	8.8%
Health Care Facilities	1.4	6.3%
Educational Institutions	1.4	6.3%
Utilities/Communications	1.1	5.1%
Distribution/Warehousing	.92	4.2%
Hotel/Motel/Resort		3.3%
Transportation	.29	1.4%
Other	.87	4.0%
TOTAL	$21.7 billion	100%
GRAND TOTAL	**$35.5 BILLION**	

[1]*Sourcebook of Criminal Justice Statistics, 1981 U.S. Department of Justice, 1982*
[2]*"Key Market Coverage," Security World* 1981
[3]Note the absence of residential, a major user of locks, alarms, fencing, and security patrols.

One of the best projections for the decade from 1980-1990 is also contained in *Private Security and Police in America*. See Table 17.3.

Two major conclusions seem obvious about the future of protective services. First, private security employment opportunities and overall expenditures for private security will continue to grow. Second, public sector employment and expenditures will grow, but at a rate much slower than private security. The results will be greater reliance on the private sector in the overall effort for crime prevention and good employment opportunities for persons seeking careers in private security.

Table 17-3
Projected Growth in Protective Service Workers

	1980	1990	Ten Year Increase
PUBLIC SECTOR			
State and Local			
Police Officers	92,981	108,642	16.8%
Patrolmen	92,972	458,922	16.8%
Sheriffs	22,276	26,601	16.8%
Police Detectives	42,705	49,913	16.8%
	550,504	643,438	16.8%
Parking Enforcement Officers	7,379	8,653	17.3%
Guards and Doorkeepers	25,170	42,428	68.6%
TOTAL STATE & LOCAL	563,053	694,619	23.4%
Federal			
Police Officers	9,179	9,905	7.9%
Police Detectives	20,635	22,267	7.9%
Guards and Doorkeepers	8,987	9,608	7.9%
All Other Workers	1,825	1,969	7.9%
TOTAL FEDERAL	40,356	43,839	7.9%
TOTAL PUBLIC SECTOR	623,409	738,438	18.4%
PRIVATE SECTOR			
Guards			
Proprietary	271,308	369,964	36.4%
Contract	341,102	443,594	30.0%
Store Detectives	18,279	27,365	49.7%
Fitting Room Checkers	8,864	11,790	33.0%
Security Checkers	230	260	13.0%
Railroad Police	2,165	1,944	[10.2%]
All Other Workers	2,395	3,976	[7.9%]
TOTAL PRIVATE SECTOR	644,343	858,893	33.3%

Source: *National Industry-Occupation Matrix,* 1980-1990, Bureau of Labor Statistics, 1982.

Professionalism

Another factor in the evolution from an occupation to a profession is a viable "Code of Ethics" for practitioners. The Private Security Task Force recognized this, as did the Private Security Advisory Council to the Law Enforcement Assistance Administration. Working cooperatively, and in concert with private security professional associations, two codes were developed, one for management and one for employees. The codes are:

Code of Ethics for Private Security Management

As managers of private security functions and employees, we pledge:

I To recognize that our principal responsibilities are, in the service of our organizations and clients, to protect life and property as well as to prevent and reduce crime against our business, industry, or other organizations and institutions; and in the public interest, to uphold the law and to respect the constitutional rights of all persons.

II To be guided by a sense of integrity, honor, justice and morality in the conduct of business; in all personnel matters; in relationships with government agencies, clients, and employers; and in responsibilities to the general public.

III To strive faithfully to render security services of the highest quality and to work continuously to improve our knowledge and skills and thereby improve the overall effectiveness of private security.

IV To uphold the trust of our employers, our clients, and the public by performing our functions within the law, not ordering or condoning violations of law, and ensuring that our security personnel conduct their assigned duties lawfully and with proper regard for the rights of others.

V To respect the reputation and practice of others in private security, but to expose to the proper authorities any conduct that is unethical or unlawful.

VI To apply uniform and equitable standards of employment in recruiting and selecting personnel regardless of race, creed, color, sex, or age, and in providing salaries commensurate with job responsibilities and with training, education, and experience.

VII To cooperate with recognized and responsible law enforcement and other criminal justice agencies; to comply with security licensing and registration laws and other statutory requirements that pertain to our business.

VIII To respect and protect the confidential and privileged information of employers and clients beyond the term of our employment, except where their interests are contrary to law or to this Code of Ethics.

IX To maintain a professional posture in all business relationships with employers and clients, with others in the private security field, and with members of other professions; and to insist that our personnel adhere to the highest standards of professional conduct.

X To encourage the professional advancement of our personnel by assisting them to acquire appropriate security knowledge, education, and training.

Code of Ethics for Private Security Employees

In recognition of the significant contribution of private security to crime prevention and reduction, as a private security employee, I pledge:

- I To accept the responsibilities and fulfill the obligations of my role: protecting life and property; preventing and reducing crimes against my employer's business, or other organizations and institutions to which I am assigned; upholding the law; and respecting the constitutional rights of all persons.
- II To conduct myself with honesty and integrity and to adhere to the highest moral principles in the performance of my security duties.
- III To be faithful, diligent, and dependable in discharging my duties, and to uphold at all times the laws, policies, and procedures that protect the rights of others.
- IV To observe the precepts of truth, accuracy and prudence, without allowing personal feelings, prejudices, animosities or friendships to influence my judgments.
- V To report to my superiors, without hesitation, any violation of the law or of my employer's or client's regulations.
- VI To respect and protect the confidential and privileged information of my employer or client beyond the term of my employment, except where their interests are contrary to law or to this Code of Ethics.
- VII To cooperate with all recognized and responsible law enforcement and government agencies in matters within their jurisdiction.
- VIII To accept no compensation, commission, gratuity, or other advantage without the knowledge and consent of my employer.
- IX To conduct myself professionally at all times, and to perform my duties in a manner that reflects credit upon myself, my employer, and private security.
- X To strive continually to improve my performance by seeking training and educational opportunities that will better prepare me for my private security duties.

Privatization — A Concept for the Future[1]

One of the unique contributions of *The Hallcrest Report: Private Security and Police in America* was the section on "Blueprint for Action" and specifically the findings, conclusions, and recommendations regarding privatization.

Private Security Contribution to Crime Prevention and Control

Research material presented in this project demonstrates the complex and far-reaching scope of private security programs in business, industry and institutions and the utilization and growth of a broad range of purchased security goods and services. Based on the sheer preponderance of evidence presented, it is clear that private security makes a sizable contribution to crime prevention and control. Crime, however, is just a part of a broad range of threats addressed by the loss prevention programs of private security, including fires, accidents, information security, materials movement, etc. Law enforcement has become increasingly aware of the presence of private security, but the substantial impact of proprietary and contract security on the overall safety and security of their communities has not been fully recognized by law enforcement administrators or operational personnel. Law enforcement executives rate the overall contribution of private security and the reduction of direct dollar crime losses by private security as only somewhat effective. They see private security's contribution to reducing the volume of crime, apprehending criminal suspects, and maintaining order as ineffective. These assessments are influenced to a great extent by law enforcement's poor ratings of private security in ten areas of performance. Law enforcement's low opinion of private security in most areas is perceived accurately by private security managers and employees: less than one-fourth of security employees think police officers view them as even performing a valuable service.

One major contribution of security personnel is their integral part in assets protection and loss prevention programs which sustain the viability and profitability of companies. In manufacturing, for example, guards prevent goods from being stolen, from fires and other forms of loss which affect the profitability of the company. Guards also protect raw materials, precious metals, production machinery and proprietary information — all of which have a direct bearing on the ability of the company to produce new goods at a profit. Private security personnel thus stimulate the economy.

Shift to Private Protection Resources

The origins of modern policing have their roots in private policing or security initiatives of the early 19th Century when there were few paid police compared to thousands of watchmen. The societal mission or role of prevention and control of crime gradually became associated with public law enforcement. The growth of modern policing and its expansion through the 1960's resulted from the redistribution of private property protection responsibilities to the public sector.

The current research documents a stabilization, and often a decline, of public law enforcement resources in recent years and simultaneously notes the growth of all segments of both proprietary and contract security. Some law enforcement administrators recognize the dramatic growth of private security in the past decade, but seem to feel that this growth is a result of the failure of law enforcement and criminal justice to do its job. In other words, if law enforcement were given adequate resources, there would be no need for widespread use of private security. Instead, these law enforcement executives see an erosion of their "turf" to private security. Hallcrest views the recent decline in law enforcement resources, increased use of private security, and increased citizen involvement in crime prevention programs as signs of a return (a century later) of the primary responsibility for protection to the private sector.

The private sector will begin bearing more of the burden for crime prevention, while law enforcement will narrow the focus of police services to crime control. Thus, we view law enforcement as assuming an increasingly reactive role even as it has expanded crime prevention programs in recent years. Hard economic realities and strained property tax bases will force law enforcement agencies to seek alternative ways to reduce their workloads. The traditionally proactive orientation of private security is well suited to assuming the non-crime-related police workload. In this research project, proprietary and contract security managers indicate a willingness to accept more responsibility for criminal incidents occurring on property being protected by them, e.g., burglar alarm response, completion of misdemeanor incident reports, and preliminary investigation. In general, law enforcement administrators are open to discussing the transfer of responsibility for criminal incidents occurring on property protected by private security, and also identified a number of police tasks as "potentially more cost effectively performed by private security." Contract security companies have expressed an interest in contracting for these non-crime-related police services, e.g., public building security, parking enforcement, and court security. Many of the activities were listed as potential areas of business growth in the next five years by national and regional contract security firms; some firms currently perform some of the candidate activities.

Smaller law enforcement agencies most affected by budget cuts and departments noting a decline or stabilization of resources are most receptive to transfer of police activities. Industry (which is frequently located in smaller communities with limited public safety services) may be willing to play a greater role in protection of its facilities, especially if tax relief (property or corporate) is involved. The greatest law enforcement interest is in transfer of burglar alarm response: nearly 70% of large police and sheriffs' departments in jurisdictions of over 500,000 population wish to be relieved of the "false alarm burden" on police workload.

It is clear that law enforcement workload could be significantly reduced—and redirected more toward "street" crime—and that the dynamics and structure of protective services delivery would be greatly changed by a realignment of public and private protection responsibilities. As long as law enforcement maintains the posture that they should bear the primary burden for protection of the community, then creative alternative solutions will be limited in the midst of dwindling public resources.

Recommendations

- **Strategic Planning for Transfer of Selected Police Activities and Contracting of Noncrime Police Services to Private Security.** The interests of the public may be best served through constructive dialogue and creative planning by law enforcement and private security to facilitate transfer of minor criminal incident responsibility and contracting of certain noncrime activities. Energy wasted on debating the quality, performance and contribution of private security could be better utilized to identify areas for contracting out, to research required legal mechanisms, and to develop tightly prescribed contract specifications of performance. The dynamics of supply and demand in the marketplace will produce a sufficient number of qualified firms, independent of any stimulus from regulation or licensing.

  This research effort has indicated that some private security personnel currently have salaries comparable to some police officers as well as substantial training and experience. Contract security company business practices and standards for security personnel would be a paramount issue in the consideration of these alternatives by government.

- **Alternative Policing Arrangements in Community Planning and Development Processes.** In addition to the transfer of responsibility for minor criminal activity on private property to private security, a broader range of linkages of private security and police services should be explored. Well-defined and homogeneous commercial and industrial districts and residential developments, developers, property owners and residents should have an opportunity to "broker" the mix of protective services which bests suits their protection needs and ability to pay. Private patrol services, for example, might be permitted to respond to certain citizen calls for service that were routed to the security officer through the technical support of the police communications center.

  Similar support might be provided to a volunteer citizen patrol trained and supervised by an area or zone police supervisor. Police administrators, themselves, might become "brokers" of policing service throughout the community, negotiating a variety of public and private protective arrangements in different areas of the community on a cost-effective basis. These efforts could be a logical extension of the progress of many law enforcement agencies in securing community involvement in the crime prevention process through Neighborhood Watch and Citizen Alert groups.

- **Police Involvement in Community Growth Planning.** While the fire services component of public safety has a long record of proactive involvement in the zoning and subdivision approval process of local government, law enforcement agencies have traditionally had little involvement in these processes. Police planners, crime prevention personnel and experienced security consultants could contribute to the review process of city and county planning and zoning departments. These activities would include recommending Crime Prevention Through Environmental Design (CPTED) concepts for individual buildings and small subdivisions, and examining larger developments for potential impacts on police services and needs for private security human and technological resources. This is done presently in some departments on a limited scale.

  The concept envisioned here, however, would also include (1) imposition of certain standards (e.g., requiring monitored alarm services and/or private security patrols for certain densities and types of commercial developments), (2) integrating security and police services in planned urban developments (PUD), and (3) facilitating special assessment or taxing districts with needs for greater or lesser levels of police services for funding both police and security services at desired levels.

- **Special Police Officer Status of Private Security.** With special police officer status, a majority of minor criminal incidents can be resolved by security personnel prior to police involvement. Establishment of preemptive state statutes on special police officer powers would allow standardized training and certification requirements to be developed, thus assuring uniformity and precluding arbitrary use of special police and deputization powers for security personnel.

  In the Baltimore case study site, many of the retailers and some of the industrial security operations opted to have certain of their security personnel designated as special police officers. The State of New York has a similar provision (for proprietary security only) and requires the security personnel to complete an approved training curriculum. In New York City, for example, some retailers utilize a 35-hour SPO training program sponsored by the Security Management Institute of John Jay College of Criminal Justice. Police officers no longer have to perform tasks of apprehension, prisoner transport, report writing, evidence presentation and court testimony for the large volume of shoplifting, trespassing, vandalism and other criminal offenses against these major retailers who use SPOs.

In such situations, the private business, rather than the general public, would bear the expense for certain police services required on its property.

- **Local and Federal Security Expenditure Tax Credits.** To enhance national crime prevention efforts, continued efforts should be directed toward enactment of a federal tax credit for certain security expenditures. One of the recent attempts at federal legislation, cited in this research, made provision for a direct offset to taxes similar to the energy tax credit. If, in fact, there is validity to the crime prevention literature supplied to the public and the deterrent capabilities of certain security technology (e.g., alarm systems), then investments in security hardware in time could result in reduced police workload. Unless the tax credit was at least $500, there is not much incentive to purchase reliable and sophisticated alarm systems or locking systems. On the other hand, if the tax credit is too low it could encourage the purchase of systems ill-suited for particular security applications, and this, in turn, could exacerbate the false alarm problem.

  Efforts should also be directed at the state and local level to reduce corporate and property tax for significant expenditures on security goods and services which offset the need for additional police services. Within the context of alternative policing arrangements discussed above, companies and organizations should have the opportunity to broker a specified level of public and private services with which it is satisfied. If the alternative arrangement reduces the cost for and burden on public police, then some offset to taxes should be allowed.

- **Activities Requiring Police Authority.** While studies on police workload have consistently shown that about 80% of police work is non-crime-related, there has never been an empirical examination of which police activities actually require the sworn authority of a police officer with his/her accompanying levels of training and skill. With local budgetary constraints forcing many law enforcement executives to practice "cut-back management" and make hard choices about the types and levels of service to be provided, attention should be focused on defining nonessential tasks — especially with private security as a viable alternative for many non-crime-related tasks. The greatest improvement in police resource efficiency will occur when sworn personnel are performing only those activities which they are uniquely qualified to perform or which could not be performed on a lower unit cost basis by the private sector with the same level of community satisfaction.

- **Alternative Modes of Policing.** An assessment should be made of (1) the basic police services the public is willing to support financially, (2) the types of police tasks/activities most acceptable to police administrators and the public for transfer to the private sector, and (3) which tasks/activities might be performed on a lower unit cost basis by the private sector with the same level of community satisfaction. An analysis should then be conducted of the organizational, environmental and legal dynamics of public and private linkages in community protection. The alternative modes or linkages with public police services should include, but not be limited to, contract security, proprietary security forces, contracts for limited police services, use of special police officer status, private developer and property management companies, and residential, neighborhood and citizen associations. For the latter group, the variables of "self-help" programs may help identify the key determinants of public willingness to assume greater responsibility for their own protection by undertaking traditional policing tasks. Relationships between cost, quality and effectiveness need to be explored for various activities and services and alternative delivery modes. Points of resistance by law enforcement administrators, government officials, private business and organizations and citizens also need to be examined.

The National Institute of Justice prepared, in 1985, a video tape titled "Corrections and the Private Sector: Fad or Future" which offers the facts for informed decision making. At the present time there appears to be a trend toward more use of private resources in the field of corrections which had strong advocates and strong objectors among sheriffs across the nation.

One of the authors of this text, Dr. Clifford W. Van Meter, highlighted the opportunities and legal cautions, as they relate in Illinois, in a speech, November 6, 1986, to 250 public and private security managers at a program sponsored by the Public/Private Liaison Committee of the Illinois Association of Chiefs of Police. Van Meter listed twelve activities that should be reviewed jointly by the public and private sectors for possible transfer, or greater interaction, in the interest of economical and efficient law enforcement.

1) Noncriminal Police Activities
2) Transportation of Prisoners
3) Burglar Alarm Response
4) School Crossing Guards
5) Animal Control
6) Funeral Escorts
7) Bank Deposit Escorts

8) Policing Special Events
9) Parking Enforcement
10) Non-injury Accident Investigation
11) Traffic Control
12) Court Security

While there are many opportunities for privatization there are also legal issues that must be reviewed. For example, in Illinois there is an Illinois Attorney General's Opinion, issued in 1982, which places many constraints on public agencies in terms of hiring private security for government service. Two key legal issues, in Illinois, based on the Attorney General's Opinion, as reviewed by Kevin McClain, Legal Counsel, Illinois Local Governmental Law Enforcement Officers Training Board are: (1) Municipalities *cannot* contract away to non-governmental entities those governmental powers granted to the municipality by the State, and (2) peace officer status is not "transferred" to private security guards. The security guard's power to arrest is limited to the powers possessed by private persons. The Attorney General's Opinion cited over twenty Illinois statutory and constitutional provisions and court decisions in support of the Opinion.

At the Michigan State University Golden Jubilee for the School of Criminal Justice in 1985, three speakers commented on the issue of use of private security in the future.

Mr. Don Bennett, Director of Security, Michigan National Bank, stated that Michigan has a joint committee of public and private security and urged "that police set up several demonstration programs to use private security for police services."

Chief William Hegarty, Grand Rapids, Michigan Police Department, stated he saw a trend toward neighborhood associations incorporating and contracting for protection services from police or private security.

Mr. Wayne Hall, Director of Security, Ford Motor Company, remarked that he saw "increasing private sector activity in the criminal justice system through privatization." He feels this concept will be expanded.

These specific references to public forums in Illinois and Michigan that presented the issue of privatization to influential practitioners may well reflect a national trend toward open and progressive discussions on the greater utilization of private security resources for crime prevention.

Professional Associations

Airport Security Council, PO Box 30705, JF Kennedy Intl. Airport, Jamaica, NY 11430

American Society for Industrial Security, 1655 N. Ft. Myer Dr., Suite 1200, Arlington, VA 22209

Associated Locksmiths of America, 3003 Live Oak St., Dallas, TX 75204

Association of Federal Investigators, 1612 K St., NW, Suite 506, Washington, DC 20006

Aviation Security Association of America — International, PO Box 17082, Washington, DC 20041

Central Station Electrical Protection Association, 1120 19th St., NW, Washington, DC 20036

Committee of National Security Companies, 87 Greenwich Ave., Greenwich, CT 06830

Computer Security Institute, 43 Boston Post Road, Northboro, MA 01532

International Association for Hospital Security, PO Box 637, Lombard IL 60690

International Association for Shopping Center Security, Suite 300, 2830 Clearview Place, NE, Atlanta, GA 30340-2117

International Association of Credit Card Investigators, 1620 Grant Ave., Novato, CA 94947

Jewelers Security Alliance of the US, Six E. 45th St., New York, NY 10017

National Burglar and Fire Alarm Association, 1120 19th St., NW, Washington, DC 20036

National Council of Investigation and Security Services, 1133 15th St., NW, Suite 620, Washington, DC 20005

National Crime Prevention Coalition, 805 15th St., NW, Room 705, Washington, DC 20005

National Crime Prevention Institute, University of Louisville, Shelby Campus, Louisville, KY 40292

National Security Industrial Association, 1015 15th St., NW, Suite 901, Washington, DC 20005

World Association of Detectives, PO Box 5068, San Mateo, CA 94402

Certified Protection Professional (CPP)

In 1972 the American Society for Industrial Security established the ASIS Institute. The combined efforts of the Society and the Institute resulted in the development of a Professional Certification Board which controls the Certified Protection Professional (CPP) program of the American Society for Industrial Security. Although this is a recent move in the professionalization of private security, it is a significant move.

By the 1980's, over 1,800 persons had received certification. The program continued to grow rapidly through August 29, 1986 (last figures available at time of printing) when the total number of CPPs by review and examination was 3,637. Of this total, 913 were certified by review, which occurred during the first six months of the certification program.[2]

Examinations are conducted almost monthly at various locations throughout the United States which provides opportunities for private security personnel to qualify for the program.

Basically, the program involves two components. Applicants must meet established criteria for certification and successfully complete a written examination. The criteria for certification are:

Criteria for Certification

An applicant must meet the following basic standards to qualify for certification.

- Experience and Education

  Experience and education prerequisites for taking the examination are shown below. Qualifying experience must be as a practitioner in the protection of assets in the public or private sector, with at least half of the experience in "responsible charge" of a security function.

No Degree	and	10 years experience
Associate Degree	and	8 years experience
Bachelor's Degree	and	5 years experience
Master's Degree	and	4 years experience
Doctoral Degree	and	3 years experience

- Affirmation of adherence to the CPP Code of Professional Responsibility
- Endorsement by a member of the Professional Certification Board, or by a person already certified as a Protection Professional
- Examination

  Achievement of a passing grade in a written examination.

Examination Content

The examination consists of objective, multiple-choice questions. It is a one-day examination divided into two parts. The mandatory or Common Knowledge examination contains 200 questions covering basic knowledge applicable in the field of security and loss prevention. The specialty examinations test knowledge through four optional 25-question examinations, chosen by the candidate from fifteen subjects on security and loss prevention practice in special areas. The *Mandatory* Subjects are:

Mandatory Subjects

1. Emergency Planning
2. Investigations
3. Legal Aspects
4. Personnel Security

5. Physical Security
6. Protection of Sensitive Information
7. Security Management
8. Substance Abuse

Optional Subjects

1. Banking & Financial Institutions
2. Computer Security
3. Credit Card Security
4. Department of Defense Industrial Security Program Requirements
5. Educational Institutions Security
6. Fire Resources Management
7. Health Care Institutions Security
8. Manufacturing Security
9. Nuclear Power Security
10. Public Utility Security
11. Restaurant & Lodging Security
12. Retail Security
13. Transportation & Cargo Security
14. Oil & Gas Industrial Security
15. Telephone/Telecommunications Security

Each question has five alternate responses, only one of which is correct. The score equals the total number of correct responses. It is to the candidate's advantage to answer each queston, even if he is not sure of the correct response. No credit is given for questions for which more than one answer is indicated.

As this program grows, it is reasonable to assume that employers of private security professionals will require this certification as a desirable attribute for applicants.

Conclusions

This text has presented material on a wide range of subjects from security lighting to locks to personnel selection. Thus, conclusions in the traditional sense cannot be made. However, the authors believe that it will have served an extremely useful purpose if the readers are encouraged to give serious consideration to pursuing private security careers and continue their efforts to obtain experience and education in the field. If this text raises more questions than answers, it will have served a useful purpose in furthering the professionalization of private security.

Discussion Questions

1. Discuss the probable future growth of private security.
2. Outline the criteria to be met for the Certified Protection Professional Program (CPP).
3. What activities best lend themselves toward privatization?

Notes

1. W.C. Cunningham, T. Taylor *The Hallcrest Report: Private Security and Police in America*, Chancellor Press (McLean, VA) pp. 246-248.
2. Lapides, Gary A. *CPP News Notes*, America Society for Industrial Security (Arlington, VA), October, 1986, No. 4.

Appendix A
Selections from the Private Security Task Force Report

Standard 1.8
Minimum Preemployment Screening Qualifications

The following minimum preemployment screening qualifications should be established for private security personnel:

1. **Minimum age of 18;**
2. **High school diploma or equivalent written examination;**
3. **Written examination to determine the ability to understand and perform duties assigned;**
4. **No record of conviction, as stated in Standard 1.7;**
5. **Minimum physical standards:**
   a. **Armed personnel—vision correctible to 20/20 (Snellen) in each eye and capable of hearing ordinary conversation at a distance of 10 feet with each ear without benefit of hearing aid;**
   b. **Others—no physical defects that would hinder job performance.**

Commentary

In order to improve the effectiveness of private security personnel, minimum preemployment screening qualifications should be established. At present, criteria for employment vary among employers, if they exist at all. This standard proposes a set of criteria that can be used by all private security employers in their preemployment screening.

The qualifications suggested are minimum. Certain employers may wish to establish stricter criteria, depending on the nature of assignment. Also, the qualifications are directed to operational personnel and generally would be inappropriate for supervisors, managers, and other specialized personnel whose duties would require more advanced knowledge and/or experience.

Age Requirements

A minimum age of 18 is recommended for all personnel. Public law enforcement agencies have constantly been hampered in recruitment by the lack of opportunity to employ sworn personnel immediately upon completion of high school. Likewise, the private security industry should not restrict itself from obtaining qualified personnel by setting unrealistic minimum or maximum ages. Many individuals are capable of performing as efficiently at age 18 as at age 21. The military services, for example, have effectively used personnel in security positions under the age of 21 for many years.

Because the establishment of career paths is an important need in the industry, age requirements need to be low enough to attract qualified applicants before they are committed to other careers. It is likely that an individual reaching age 21 would have already identified career aspirations, and a job in private security would, at best, be only a secondary interest. As mentioned previously in Goal 1.1, personnel will function more effectively when they are performing the job they want to do.

Educational Requirements

The *RAND Report* (Vol. 1) stated that, in response to a survey questionnaire, two-thirds of the regulatory agencies indicated that minimum educational requirements should be mandatory for private security personnel. Of the two-thirds favoring minimum educational requirements, one-third indicated that private security personnel should be high school graduates. Others thought education beyond high school would be a more appropriate requirement for some categories. For example, two recommended college education for investigators; two proposed polygraph-school graduation for lie-detection examiners; one believed that supervisors should have some college training. Significantly, one-third of the survey respondents thought no minimum educational requirements should be established.

For the purpose of this standard, educational requirements are classified in two main categories: (1) basic educational qualifications and (2) ability to understand and perform duties assigned. The basic educational qualifications can be met by a high school diploma or an equivalent written examination designed to measure basic educational aptitudes. The employer should be careful, however, to utilize only those tests that have been proven valid and reliable.

The second educational requirement—the ability to understand and perform duties assigned—is determined through a written examination. Here, again, the employer should use only validated tests. Furthermore, there should be a close cause-effect relationship between the tests and the job description in accordance with the following Equal Opportunity Employment Commission guideline on employment testing procedures:

> The Commission accordingly interprets "professionally developed ability test" to mean a test which fairly measures the knowledge or skills required by the particular job or class of jobs which the applicant seeks, or which fairly affords the employer a chance to measure the applicant's ability to perform a particular job or class of jobs. The fact that a test was prepared by an individual or organization claiming expertise in test preparation does not, without more, justify its use within the meaning of Title VII (Civil Rights Act of 1964).

The two categories of educational requirements are not mutually inclusive or exclusive. For example, a high school graduate might have psychological characteristics that would indicate this person should not be armed. Conversely, a person who did not graduate from high school but passes the equivalent written examination might be found to be psychologically qualified to carry a weapon. A high school diploma, in and of itself, should not necessarily be a prerequisite for armed personnel, but regulatory agencies, for administrative reasons, may set such a requirement.

The National Advisory Committee on Criminal Justice Standards and Goals (NAC) did not agree with the Task Force's position that a high school diploma or equivalent written examination should be a minimum preemployment screening qualification for all private security personnel. Although agreeing that this requirement was appropriate for armed guards and certain security activities, the NAC believed that a written examination to determine if an individual had the ability to understand and perform the duties involved was adequate for other security assignments. The NAC believed that individuals who were competent to perform these other security assignments would be denied employment in the field if the high school education level requirement was a minimum standard. However, it is the opinion of the Private Security Task Force that the basic knowledge engendered by a high school diploma (or equivalent written examination) is important for emergency situations that may arise. The written ability examination may not test for those skills outside the private security employee's job description, and he may, therefore, not be able to handle the emergency situation. Also, in the furtherance of the development of a professional private security industry, high-school-level education is considered necessary in the judgement of the Task Force.

Conviction Records

Conviction records, except for certain minor offenses, should preclude private security employment. Standard 1.7 discusses this topic fully and points out the responsibilities assumed by private security personnel to the public and to the role of crime prevention. For the public to have confidence in private security personnel, employers should select persons of high moral integrity. In order to facilitate implementation of this standard, this report calls for the cooperation of government agencies in supplying pertinent conviction records.

Physical Requirements

Physical requirements should not be unnecessarily restrictive. In most cases, specific physical qualifications, such as height and weight, would be inappropriate. The results of a study released by the International Association of Chiefs of Police and the Police Foundation, and published in [the] Dec. 1, 1975, issue of *Crime Control Digest*, confirm that height requirements, for example, have little relation to performance and tend to unnecessarily reduce the available pool of qualified applicants:

> The authors... say that they found no data, either from their survey of five police departments or from their search of literature on the subject, that show that the height of a police officer does affect performance....
>
> ...Height requirements can vastly reduce the pool of applicants who have personal qualities needed by police departments.
>
> For example, 56 percent of young adult males and 99 percent of young adult females would be excluded from employment by a minimum height requirement of 5 feet 9 inches.

Although the authors of this study were hampered in their research by the lack of a large comparative population, the results point the way to a selection system without height requirements.

However, private security employers should not totally disregard physical standards or take them lightly. One employer cited in the *RAND Report* (Vol. 1) said, "Some standards are a joke. While we require a physical exam for employment, if the man can take three steps he passes the physical." In general, physical requirements should be determined by the nature of the job the applicant would be performing. Any physical defect that would interfere with ability to perform assigned duties would disqualify the applicant.

Differentiation should be made [between] physical qualifications for armed personnel and others. Obviously, good eyesight and hearing are vital to anyone who carries a weapon; therefore, specific vision and hearing qualifications should be established for armed private security personnel in consideration of protecting both themselves and the public.

Standard 2.5
Preassignment and Basic Training

Any person employed as an investigator or detective, guard or watchman, armored car personnel or armed courier, alarm system installer or servicer, or alarm respondent, including those presently employed and part-time personnel, should successfully:

1. Complete a minimum of 8 hours formal preassignment training;

2. Complete a basic training course of a minimum of 32 hours within 3 months of assignment. A maximum of 16 hours can be supervised on-the-job training.

Commentary

Other standards have highlighted the lack of training in the private security industry. This lack has inspired much criticism, most of it directed specifically at the failure of the industry to properly and adequately prepare its operational-level personnel. *The Other Police*, a report on the Ohio private security industry, contains a section entitled "Training: Infrequent, Incomplete, and Misdirected." It points out that fewer than 25 percent of Ohio's guards hold training certificates from the Ohio Peace Officer Training Council and that training throughout the State is decreasing instead of increasing.

Lack of private security training also tends to generate friction with public law enforcement agencies. For example, law enforcement officers, working in the same community with private security guards, investigators, and so forth, often look down on their abilities and question their judgments, because private security personnel are untrained. The public law enforcement has made tremendous progress in the past decade in both adequacy of training and quality of courses, but the private security industry has barely taken a step in this direction.

A survey of members of the American Society for Industrial Security (ASIS) indicated a present range of 4 to 80 hours of training for newly hired personnel. Table 2.2, from the *RAND Report* (Vol. II), further illustrates the wide range and general inadequacy of initial training in a sample of 11 private security companies.

Other findings from the *RAND Report* (Vol. II) indicate that a large percentage of private security guards do not know their legal powers to detain, arrest, search, or use force. Frequently, in fact, they lack understanding of the basic policies and procedures of their functions. The following comment from a former guard, who was beaten during a robbery, vividly illustrates the need for additional training:

> For $1.60 per hour I wouldn't stick my neck out again. Anybody who does is crazy. I stand around looking cute in my uniform. Don't let

anybody tell you a guard doesn't need training. If I'd had it I might had known what the hell was going on.

Private security professionals do recognize the importance of training. The previously mentioned ASIS survey revealed that 76 percent of the respondents believed training standards were "very important"; 15 percent, "somewhat important"; and 1 percent, "not important." Yet, until specific standards are required, private security training is not likely to improve.

Preassignment Training

This standard recommends that training requirements be initiated for all operational private security personnel. A RAND survey of private security personnel in California revealed that 65 percent of the respondents had received no training prior to beginning work. Because the instruction received at this stage familiarizes the employee with the responsibilities of the job and establishes certain basic skills and concepts, it is recommended that every private security employee successfully complete 8 hours of preassignment training before commencing work.

Due to the complexity of functions performed by private security personnel, the final determination of subject content for preassignment training will need to be made by employers and regulatory agencies; however, the following topical outline is recommended as a general guide. It is based on a model originally prepared by the Private Security Advisory Council, included in their *Model Private Security Licensing and Regulatory Statute*, and designed for guards. Obviously, some additions in content were necessary to expand it to meet the broader spectrum of personnel included in this standard.

Private Security 8-Hour Preassignment Training Course

Section I—Orientation: 2 hours that include the following topics:

- What is security?
- Public relations.
- Deportment.
- Appearance.
- Maintenance and safeguarding of uniform and/or equipment.
- Notetaking/Reporting.
- Role of public law enforcement.

Section II—Legal Powers and Limitations: 2 hours that include the following topics:

- Prevention versus apprehension.
- Use of force.

- Search and seizure.
- Arrest powers.

Section III—Handling Emergencies: 2 hours that should include appropriate topics pertinent to the job functions to be performed by the employee:

- Crimes in progress.
- Procedures for bomb threats.
- Procedures during fires, explosions, floods, riots, and so forth.
- Responding to alarms.

Section IV—General Duties: 2 hours that should include the appropriate topics pertinent to the job functions to be performed by the employee:

- Fire prevention and control.
- Inspections.
- Interviewing techniques.
- Patrol.
- Safeguarding valuable property.
- Safety.
- Surveillance.

The following model preassignment training programs are intended to explain how the program could be implemented for guards or watchmen or alarm respondents. Again, specific recommendations are not established because of the complexity of training needs, but the outline may prove helpful as a general guideline. The hour designation used in all training standards is a 50-minute block of instruction that is standard for training and education curriculums.

Model Preassignment Training Program for a Guard or Watchman

Section I—Orientation (2 hours)	**Minutes**
• What is security?	15
• Public relations	15
• Deportment	15
• Appearance	10
• Maintenance and safeguarding of uniforms and/or equipment	20
• Notetaking/Reporting	15
• Role of public law enforcement	10
Section II—Legal Powers and Limitations (2 hours)	
• Prevention versus apprehension	40
• Use of force	25
• Search and seizure	15
• Arrest powers	20

Table 2.2. Current Private Security Guard Training Programs

	Initial Prework Training								Initial On-the-Job Training				
Program	Talking with Super-visors (hours)	Read Manual	View Films/ Slides (hours)	Class (hours)	Test	Firearms Range	Trained on Previous Job	Total (hours)	By Super-visor (hours)	By Fellow Employee (hours)	Written Post Orders	Total (hours)	Total Initial Training (hours)
Company A: Small Contract Guard Firm	1/2 to 1	None	None	None	None	N/A	None	1/2 to 1	8 to 16	None	Yes	8 to 16	8 1/2 to 17
Company B: Small Contract Guard Firm	1 to 2	Yes	None	None	Yes	Yes	None	2 1/2 to 3 1/2	8 to 16	None	Yes	8 to 16	10 1/2 to 19 1/2
Company C: Medium Contract Guard Firm	1 to 3	Yes	1 1/2	None	Yes	Yes	None	5 to 7	8 to 16	None	Yes	8 to 16	13 to 23
Company D: Large Contract Guard Firm (full- and part-time)	1 to 2	Yes	2	None	Yes	Yes	None	6 1/2 to 7 1/2	1 to 8	None	Yes	1 to 8	7 1/2 to 15 1/2
Company E: Large Contract Premium Guard Firm	1 to 2	Yes	2	40 to 80	Yes	Yes	None	46 1/2 to 87 1/2	1 to 8	None	Yes	1 to 8	47 1/2 to 95 1/2
Company F: Large Contract Guard Firm													
a. Regular	None	Yes	1	9	None	Yes	None	12	1 to 8	None	Yes	1 to 8	13 to 20
b. Temporary	3 to 4	None	1	None	None	None	None	4 to 5	1/2	None	None	1/2	4 1/2 to 5 1/2

Company G: Large Contract Guard Firm													
a. Regular	None	Yes	None	10	Yes	Yes	None	11	1/2 to 1	None	Yes	1/2 to 1	10 1/2 to 11
b. Temporary	None	None	None	8	None	None	None	8	1/2	None	None	1/2	8 1/2
Company H: Small Contract Patrol Guard Firm	1 to 2	None	None	None	None	Yes	None	3 to 4	16	None	Yes	16	19 to 20
Company I: Inhouse Guards (Bank)	2 to 4	Yes	None	None	None	Yes	Occa-sionally	5 to 7	80 to 120	None	Yes	80 to 120	85 to 127
Company J: Inhouse Guards (Research)	1 to 4	Yes	None	None	None	N/A	None	3 to 6	None	160	Yes	160	163 to 166
Company K: Inhouse Guards (Manufacturing)	1/2 to 1	Yes	None	None	None	N/A	Manda-tory	1/2 to 2	None	24	Yes	24	25 1/2 to 26

Source: Kakalik, James S., and Sorrel Wildhorn. *The Private Police Industry: Its Nature and Extent.* Vol. II, R-870/DOJ. Washington, D.C.: Government Printing Office, 1972, p. 33.

Section III—Handling Emergencies (2 hours)

• Procedures for bomb threats	40
• Procedures during fires, explosions, floods, riots, and so forth	60

Section IV—General Duties (2 hours)

• Patrol	40
• Fire prevention and control	30
• Safety	30

Model Preassignment Training Program for an Alarm Respondent

Section I—Orientation (2 hours)	Minutes
• What is security?	15
• Public relations	10
• Deportment	10
• Appearance	10
• Maintenance and safeguarding of uniforms and/ or equipment	30
• Notetaking/Reporting	15
• Role of public law enforcement	10
Section II—Legal Powers and Limitations (2 hours)	
• Prevention versus apprehension	25
• Use of force	25
• Search and seizure	30
• Arrest powers	20
Section III—Handling Emergencies (2 hours)	
• Crime in progress	20
• Responding to alarms	80
Section IV—General Duties (2 hours)	
• Interviewing techniques	40
• Patrol	30
• Safeguarding valuable property	30

Model Preassignment Training Program for an Armored Car Guard[1]

Section I—Orientation (2 hours)	Minutes
• Protective transportation: History of armored car industry Basic elements of service Interface with the financial community	50

- The company:

History of employer	15
Organizational structure	15
Wages and benefits	20
Driver/guard	
Messenger/guard	
Custodian/guard	

Section II—Legal Powers and Limitations (2 hours)

- Parameters of operation

We are not policemen or stationary guards	10
Theory of bailment	10
Use of selective force in defensive role	25
Weapons philosophy	25
Physical force and its operational application	20
Restraints in dissemination of confidential information	10

Section III—Handling Emergencies (2 hours)

- Emergency situations (an overview)

Defining the threat	30
Robbery	
On the sidewalk	
In customer's premises	
In the truck	
Political terrorists versus conventional criminal	70
Extortion	
Abduction	
Ambush	
Bomb threats	

Section IV—General Duties (2 hours)

• Fire procedures	25
• Traffic accidents	25
• Rules and regulations	40
Uniforms	
Equipment (familiarization)	
Armored truck	
Handtruck	
Seals and bags	
Terminals	
Vaults	
Security areas	
• Deportment	10

In implementing the suggested preassignment training programs, the following factors should be noted:

1. All topics in Sections I and II should be covered in some portion of the 2 hours assigned.

2. Only pertinent topics in Sections III and IV need to be included in the 2 hours assigned.

3. Supervised, on-the-job training cannot be used to meet preassignment training.

4. Lectures, films, programmed learning, and other training methods can be used.

Basic Training

Upon successful completion of preassignment training, the employee should be allowed to begin work, but training should not stop at this point. Additional training is needed to provide the skills, knowledge, and judgment necessary for efficient, effective job performance. Although the importance of this training cannot be overemphasized, it is recognized that the high cost of training may place a heavy economic burden on some employers. Therefore, a realistic minimum of 32 hours of basic training is recommended in addition to preassignment training. This training should be completed over a 3-month time period and may include a maximum of 16 hours on-the-job training.

Although many may believe that the 32-hour training standard is totally inadequate, it is a progressive step in terms of the amount of training presently provided. Admittedly, it is far short of the 400 hours recommended in 1973 for sworn police officers by the National Advisory Commission on Criminal Justice Standards and Goals. It should be understood, however, that Federal, State, and local tax dollars support training for public law enforcement officers, but only limited monetary resources are available to provide training for private security personnel. Ultimately, a large portion of the cost would have to be borne by the consumer. Although, in some instances, employees are required to pay the cost of their own training, this practice is discouraged unless such training is personally sought by the individual to prepare himself for private security employment. The 32-hour minimum basic requirement is believed to be economically feasible for implementation by all; those employers financially capabale of providing additional training should surpass the 32-hour minimum.

Basic training requirements, as stated in this standard, should apply to both presently employed and part-time personnel. Because of the prevalent lack of training throughout the private security industry, many present employees are not adequately prepared for the responsibilities of their positions. Thus, they should be required to have the same training as newly hired personnel if uniform quality of performance is to be achieved. Part-time employees also assume the same responsibilities and need the same amount of training.

By allowing 16 hours of the basic training to be completed on the job, employers can maximize the training effect. However, it is very important that close supervision is provided for employers to meet the intent of the standard. With appropriate supervision, an employee can effectively relate classroom instruction to the specific job performed. In this manner, training can take on added significance and reality.

Responsibility for implementation of private security basic training would rest with employers and State regulatory agencies. As with preassignment training, these persons ultimately would have to determine the actual subjects presented in basic training. However, to provide general guidance in determining curriculums, the following topical outline for a 32-hour basic course of training is offered:

Private Security 32-hour Basic Training Course

Section I—Prevention/Protection

- Patrolling.
- Checking for hazards.
- Personnel control.
- Identification systems.
- Access control.
- Fire control systems.
- Types of alarms.
- Law enforcement/Private security relationships.

Section II—Enforcement

- Surveillance.
- Techniques of searching.
- Crime scene searching.
- Handling juveniles.
- Handling mentally disturbed persons.
- Parking and traffic.
- Enforcing employee work rules/regulations.
- Observation/Description.
- Preservation of evidence.
- Criminal/Civil law.
- Interviewing techniques.

Section III—General emergency services

- First aid.
- Defensive tactics.
- Fire fighting.
- Communications.

- Crowd control.
- Crimes in progress.

Section IV—Special problems

- Escort.
- Vandalism.
- Arson.
- Burglary.
- Robbery.
- Theft.
- Drugs/Alcohol.
- Shoplifting.
- Sabotage.
- Espionage.
- Terrorism.

To allow flexibility for individual situations and yet provide reasonable controls, the following items should be considered:

1. A minimum of 4 classroom hours should be provided in each of the sections.
2. A maximum of 16 hours supervised, on-the-job training should be permissible.

The following models explain how the basic training course can be implemented:

Model 1. Maximum classroom hours

Section	Classroom hours Minimum	Maximum
• Prevention/Detection	4	16
• Enforcement	4	16
• General/Emergency services	4	16
• Special problems	4	16

Discussion: The maximum of hours in each section can be modified in any way that is appropriate to the training needs; however, 4 classroom hours should be provided in each section. For example, an alarm response runner could follow these courses:

Section	Classroom hours
• Prevention/Detection	20 or 16
• Enforcement	4 or 5
• General/Emergency services	4 or 5
• Special problems	4 or 6

(May use any combination provided a minimum of 4 classroom hours are in each section and the total hours are 32.)

Model 2. Minimum classroom hours

Section	**Classroom hours**
• Prevention/Detection	4
• Enforcement	4
• General/Emergency services	4
• Special problems	4

(Should include 16 hours of supervised on-the-job training.)

Discussion: In many cases needs can best be met by training the employee in the job setting after providing basic knowledge and skills. This model provides the necessary latitude for these situations.

Model 32-hour Basic Training Course for Armored Car Guards[2]

Section I—Prevention/Detection

(Operating procedures)—6 hours	**Minutes**
• Crew operations In the terminal On the street On customer's premises	100
• Armored truck and equipment drills	50
• Packaging	25
• Receipting system	50
• Reporting and forms preparation	25
• Police liaison	50

Section II—Enforcement (Robbery and loss)—4 hours

• Case studies of attacks on men and equipment	50
• Role playing	150

Section III—General/Emergency services

(Emergency response)—6 hours	**Minutes**
• Trauma treatment (10-minute medicine) Gunshot Explosion Burns Vehicle accidents CPR training	100
• Basic firefighting techniques	25
• Basic self-defense	75

- Bomb threats 50
  - Bomb recognition
  - Vehicle inspection
  - Tactical reaction to a bomb
  - Bomb call threat to terminal
  - Customer premises threat
  - Suspicious device located
    - On vehicle
    - In the terminal
    - In customer's premises
- Use of communications 50

Section IV—Special problems (Emergency drivers)—4 hours Minutes

- Defensive driving 40
- Philosophy of offensive driving 30
  - Counterambush
    - Urban
    - Rural
- Night driving 30
- Hands-on driver training 100

(Should include at least 12 hours of supervised on-the-job training to include examination and course evaluation.) (Note: A number of industry representatives indicated that more than 12 hours of supervised on-the-job training would be provided to meet employees'needs.)

Discussion: Because the vast majority of armored car guards are armed (and to meet the firearms training of Standard 2.6), the Training Committee of the National Armored Car Association included the following outline as part of the basic training program:

Firearms Training

- Company and industry policy on use of weapons
- Legal limitations
- Firearms safety
- Care and cleaning
- Basic revolver training
  - Combat firing
  - Use of gunports
  - Use of shotgun
  - Qualification and certification

The previous models provide the extremes of the standard. The 32 hours of training could be implemented in a variety of ways, with the following factors in mind:

1. The total basic training program encompasses 32 hours.
2. The minimum classroom hours are 16.

3. The maximum supervised on-the-job training is also 16.

4. The ratio between the minimum classroom hours and the maximum supervised, on-the-job training can vary (e.g., 20 classroom hours and 12 on-the-job training hours).

Several final points involving this training standard are offered for purposes of clarity:

1. The issue of an exemption from the requirements of this standard—a "grandfather" clause—for all private security personnel was considered and rejected because the training standard is a basic minimum and all personnel should receive it.

2. Formal or classroom training, both for preassignment and basic, can be lectures, films, slides, programmed instruction, and the use of other training media.

3. Supervised, on-the-job training means that personnel receive close observation and supervision. Merely being assigned to a job cannot be called on-the-job training.

4. The 3-month period to complete training is included to allow employers the flexibility to group personnel into training sessions that best meet the employers' and employees' needs, and also to minimize the economic losses caused by training persons who leave after a short period of time.

5. At least 1 hour for examinations should be included in the training curriculum and should be taken as a reduction in the supervised, on-the-job training hours. Depending on the delivery system, it may be advisable to have a testing block of time for each section.

6. Part-time personnel means all personnel who work less than full-time and includes personnel listed as temporary, half-time, and so forth.

7. Some may view the 8-hour preassignment training as totally inadequate preparation before starting employment. More preassignment training, as appropriate is encouraged. Many subjects in the basic course could be included in an expanded preassignment course.

As stated earlier, many security professionals would believe that the training recommended is minimal and that additional specialized training would be needed, depending on the skills, knowledge, and judgment required for certain assignments. For example, private investigators and detectives may require more training than this standard specifies. The specific amount of time and course content would have to be determined on an individual basis. The following list is presented to illustrate the types of specific subjects that could be included in the additional training:

- Background investigation
- Civil court procedures.
- Civil damage suits.
- Criminal court procedures.

- Collection and preservation of evidence.
- Crime prevention.
- Custody and control of property.
- Fingerprints.
- Followup investigations.
- Identification of persons.
- Industrial investigations.
- Insurance investigations.
- Interviews.
- Investigation and security as a professional vocation.
- Investigator's notebook.
- Mock crime scene.
- Modus operandi.
- Motion and still cameras.
- Obtaining information from witnesses.
- Plaintiff investigations.
- Preemployment investigations.
- Preliminary investigations.
- Preventive security.
- Principles of investigation.
- Purpose of private investigation.
- Report writing.
- Retail store investigation.
- Rules of evidence.
- Search and seizure.
- Sources of information.
- Surveillance and stakeout.
- Taking statements.
- Testifying in court.
- Undercover assignments.

Although many of these topics may seem more important to public law enforcement investigators, they are also relevant to private investigators. For example, many cases developed by private investigators end up in civil court while others are filed in criminal court. Thus, the training of private investigators should properly prepare them for this eventuality.

Guards or watchmen, couriers, alarm system installers or repairers, and alarm respondents may also require additional training, and similar, expanded subject outlines can be developed to provide the needed training. The use of investigators and detectives as one example should not be construed as an indication that they are the only categories of private security personnel who might need specialized training.

Selected References

1. Brennan, Dennis T. *The Other Police.* Cleveland: Governmental Research Institute, 1975.

2. Criminal Justice Institute: "88-Hour General Security Program." Detroit: Criminal Justice Institute.

3. Eversull, Kenneth S. "Training the Uniformed Officers," *Security World*, May 1967.

4. Ford, Robert E. (supervising ed.). *TIPS: A Continuous Program of Training and Information for Private Security.* Santa Cruz, Calif.: Davis Publishing Company, Inc., 1975.

5. Kakalik, James S., and Sorrel Wildhorn. *The Private Police Industry: Its Nature and Extent,* Vol. II, R-870/DOJ. Washington, D.C.: Government Printing Office, 1972.

6. Kelly, James. Address before the Private Security Advisory Council, Chicago, Ill., July 11, 1975.

7. National Advisory Commission on Criminal Justice Standards and Goals. *Police,* Washington, D.C.: Government Printing Office, 1973.

8. National Council on Crime and Delinquency. "Minimum Standards for the Training of Private Security Guards." Hackensack, N.J.: National Council on Crime and Delinquency, May 1973.

9. Norell and Acqualino. "Scarecrows in Blue," *The Washingtonian,* August 1971.

10. O'Hara, Charles E. *Fundamentals of Criminal Investigation.* 3d ed. Springfield, Ill.: Charles C. Thomas Publishers, 1973.

11. Post, Richard S. "Application of Functional Job Analysis to the Development of Curriculum Guidelines for Protective Services Field." Ph.D dissertation. Madison: University of Wisconsin, 1974.

12. Private Police Training Institute. "The Course Outline." Louisville, Ky.: Jefferson Community College, 1975.

13. Private Security Advisory Council. *Model Private Security Licensing and Regulatory Statute.* Washington, D.C.: Law Enforcement Assistance Administration, 1975.

14. Private Security Task Force. "American Society for Industrial Security (ASIS) Survey Results." (See Appendix I to this report.)

15. "Survey of Security Instruction Time." *Security World,* February 1972.

16. Vanderbosch, Charles G. *Criminal Investigation,* Gaithersburg, Md.: International Association of Chiefs of Police, 1968.

17. Wilson, O.W. *Police Administration.* 2d ed. New York: McGraw-Hill, Inc., 1963.

Notes

1. This model preassignment training program was prepared by the Training Committee of the National Armored Car Association at the request of the Private Security Task Force.
2. Prepared from model 32-hour basic training course presented to the Task Force by the Training Committee of the National Armored Car Association.

Standard 2.6
Arms Training

All armed private security personnel, including those presently employed and part-time personnel, should:

1. Be required to successfully complete a 24-hour firearms course that includes legal and policy requirements—or submit evidence of competence and proficiency—prior to assignment to a job that requires a firearm;
2. Be required to requalify at least once every 12 months with the firearm(s) they carry while performing private security duties (the requalification phase should cover legal and policy requirements).

Commentary

Armed personnel are defined as persons, uniformed or nonuniformed, who carry or use at any time any form of firearm. The serious consequences, for both employers and employees, when untrained personnel are assigned to jobs that require firearms are obvious. These consequences can be generally outlined as:

1) Self-injury because of mishandling of the weapon;
2) Injury to others, often innocent bystanders, because of lack of skill when firing the weapon; and
3) Criminal and/or civil suits against both employers and employees resulting from the above actions.

A 1974 study by the Institute for Local Self Government revealed that 45 percent of licensed California private security agency heads admitted to providing no formal preassignment instruction in firearms use, and 40 percent indicated a lack of weapons retraining. Even more revealing and disturbing, 55 percent of the employees surveyed said they sometimes carry firearms, but only 8 percent had received firearms training in their present jobs.

The *RAND Report* (Vol. II) indicated that 49 percent of private security personnel carried firearms, but only 19 percent had received any firearms training in their present jobs. The following statement from the *Philadelphia Magazine* pointedly reveals one employee's feelings:

> One guard who shot two people within two weeks in Philadelphia complained that the detective agencies were "taking young jitterbugs off the street, putting guns in their hands and giving them no training. The companies are cleaning up, man, and they ought to spend some of that money to train us."

Statistics and reports, such as the above, emphasize the vital necessity of adequate training for all personnel who are to carry firearms in their private security duties, even if they are instructed never to use them. Employers cannot ignore this need or attempt to evade it, as was done in the following example: An article in the January 1973 issue of *Police Weapons Center Bulletin* reported that a Virginia firm was manufacturing fake replicas of standard police revolvers and marketing them to security agencies for issuance to guards. According to the article, 30 private security agencies had purchased these replicas to equip their guards, thus eliminating the problem of issuing real firearms to untrained or semitrained personnel. The consequences of this action could be tragic. No firearms should ever be issued to private security personnel, unless the weapons are authentic and employees are well trained in their use and legal implications.

The intent of this standard is that employees should not be allowed to carry firearms while performing private security duties unless they can demonstrate competency and proficiency in their use. In attempting to construct an appropriate training course for firearms instruction, many existing courses were reviewed. The recommended course that follows is designed for persons armed with revolvers and may require modification for other weapons or for adaptation to local situations. Dick Mercurio, training coordinator, Southwestern Illinois Law Enforcement Commission, indicated that persons were trained in 1974 and 1975 with about a 90 percent successful completion rate by generally following this classroom outline. In general, the recommended course includes 6 hours of classroom and 18 hours of range firing.

Classroom

Topic I Legal and policy restraints—3 hours

1. Rights of private security personnel to carry weapons and powers of arrest
2. Statutory references
3. Policy restraints

Topic II Firearms safety and care and cleaning of the revolver—2 hours

1. Nomenclature and operation of the weapon
2. Performance of cartridge
3. Safety practices on duty and at home
4. Range rules
5. Care and cleaning of the weapon

Topic III Successful completion of written examination—1 hour

1. At least 20 questions on the above topics with a minimum passing score of 70 percent
2. Should be designed so that persons with other and/or prior experience can demonstrate competence in the subject areas.

Range[1]

Topic I Principles of marksmanship—2 hours

1. Shooting stance
2. Gripping and cocking the revolver
3. Sighting
4. Trigger control
5. Breathing control
6. Speeding loading and unloading techniques

Topic II Single action course—8 hours

- Distance: 25 yards
- Target: silhouette
- Rounds fired for qualification: 30
- Minimum passing score: 18 hits (60 percent)
- Stages of the course:
  1. Slow fire—consists of 10 shots fired in a total time of 5 minutes.
  2. Time fire—consists of two strings of 5 shots each. Each string is fired in a time limit of 20 seconds.
  3. Rapid fire—consists of two strings of 5 shots each. Each string is fired in a time limit of 10 seconds.
- Courses fired:
  1. Slow fire practice—30 rounds
  2. Time fire practice—6 strings—30 rounds
  3. Rapid fire practice—6 strings—30 rounds
  4. Practice course—30 rounds
  5. Record course—30 rounds

Topic III Double action course—8 hours

- Distance: as outlined below
- Target: silhouette
- Rounds fired for qualification: 72
- Minimum passing score: 43 hits (60 percent)
- Stage of the course: 7 yard line—Crouch position
  - a. First phase:
    - (1) load; draw and fire 1 and holster on the whistle command (6)
    - (2) load; draw and fire 2 and holster on the whistle command (6)
    - (3) repeat (1) and (2), using weak hand (12)
  - b. Second phase:
    - (1) strong hand—time 30 seconds—load; draw on the whistle, fire 6; reload and fire 6 more (12)
    - (2) weak hand—time 30 seconds—load; draw on the whistle, fire 6; reload and fire 6 more (12)

- Courses fired: The above courses will be fired 4 times in the following sequence:
    1. A practice course (72)
    2. Skip loading with 3 rounds each string (24)
    3. Preliminary record course (72)
    4. Firing for record (72)

The purpose of range training is to ensure that private security personnel meet minimum proficiency requirements. If, for example, a student qualifies during the preliminary or practice rounds, it may be appropriate to remove him from the range course and give the instructor more time with students who are having difficulties. However, no person should be considered proficient, and assigned to a job that requires a firearm, unless he meets the minimum qualifications outlined.

Although not specifically stated in the standard, all instructors should be qualified through the National Rifle Association or other comparable qualifications programs.

In summary, the following requirements should be stressed for personnel carrying firearms:

1. Competence in the classroom subjects (minimum score of 70 percent) and proficiency with the weapon (minimum score of 60 percent) should be met before assigning any personnel to jobs that require firearms.
2. Personnel should be trained in the use of any weapon they carry.
3. They should meet the weapon proficiency requirements at least once every 12 months.

One study, *Private Security Survey and Ordinance for St. Petersburg, Florida*, recommended a more stringent requirement for point three—retraining courses to be held at 6-month intervals.

Employers also should consider preparation of a firearms policy form, including safety rules, policies regarding discharge of weapons, and other pertinent matters. Employees would be required to sign the form every 3 or 4 months, indicating they understand the policies. Their supervisors also would be required to sign the form. This system has been used for a number of years in the military services and has been an effective reminder of firearms policy.

No amount of required training can guarantee that weapons abuses will be eliminated or that accidents will cease to occur. However, a firearms training program, as outlined, can reduce the incidence of these types of problems. The necessity of training is apparent; the risks are too great without it. The private security industry should immediately provide training for all of its armed personnel.

Selected References

1. Chapman, Samuel G., and Thompson S. Crockett. "Gunsight Dilemma: Police Firearms Policy," *Police*, March-April 1963.

2. Institute for Local Self Government. *Private Security and the Public Interest*. Berkeley, Calif.: Institute for Local Self Government, 1974.

3. International Association of Chiefs of Police. "A Questionable Practice: Security Officers 'Armed' with Fake Weapons," *PWC Bulletin*, January 1973.

4. Kakalik, James S., and Sorrel Wildhorn. *The Private Police Industry: Its Nature and Extent*, Vol. II, R—870/DOJ. Washington, D.C.: Government Printing Office, 1971.

5. Mallowe, Mike. "Willie Lee Weston Is Armed and Dangerous." *Philadelphia Magazine*, August 1975.

6. Martensen, Kai R. "Private Security Survey and Ordinance for St. Petersburg, Florida." Sunnyvale, Calif.: Public Systems Incorporated, 1975.

7. Silvarman, Allen B. I. "Firearms Training," *Security Distributing and Marketing*, December 1974.

8. Strobl, Walter M. "Private Guards: Arm Them or Not," *Security Management*, January 1973.

Notes

1. The training hours for the range may seem excessive. However, it must be remembered that many of the personnel may have had no previous firearms training. Other factors that cause delays, such as the number of shooting positions available in relation to the number of students, should also be considered. The outline for the range course was supplied by Dick Mercurio, training coordinator, Southwestern Illinois Law Enforcement Commission.

Standard 11.2
Registration Qualifications

Every applicant seeking registration to perform a specific security function in an unarmed capacity should meet the following minimum qualifications:

1. Be at least 18 years of age;

2. Be physically and mentally competent and capable of performing the specific job function being registered for;

3. Be morally responsible in the judgment of the regulatory board; and,

4. Have successfully completed the training requirements set forth in Standard 2.5

Commentary

The 1967 *Task Force Report: The Police* of the President's Commission on Law Enforcement and Administration of Justice stated that "policing a community is personal service of the highest order, requiring sterling qualities in the individual who performs it.... Few professions are so peculiarly charged with individual responsibility." Although the quote is directed toward law enforcement personnel, it is equally applicable to private security personnel, who likewise often must make instantaneous decisions affecting lives and property.

As pointed out in the preceding standard, the nature of the role of the private security industry demands that steps be taken to upgrade the quality of its personnel. Research has indicated that far too many security personnel, charged with protection of life and property, are either incompetent or of questionable character. Yet, existing personnel selection requirements and procedures do not screen out the unfit. If costly and dangerous losses both to business and society are to be prevented, measures for improvement need to be devised.

Chapter 1 of this report makes a number of recommendations for improving the quality of private security personnel. These recommendations reflect reasonable standards that should be established. However, despite the validity of the recommendations, it is recognized that certain actions may never be instituted unless mandated by law. Therefore, in order to improve the quality of security personnel, it was felt that certain minimum qualifications should be established for registration.

Recognizing the desire to attract high school graduates who might make a career in the private security field, a minimum age requirement of 18 is suggested, thus enabling businesses to compete for qualified young people. It is believed that personnel who do not possess the necessary maturity so often associated with age would not meet other requirements. No attempt is made, however, to impose a maximum age restriction. Any individual who can meet the physical and mental qualifications established by the regulatory board should be allowed to perform security functions, regardless of age.

Physical qualifications are not specifically enumerated, because each particular job function requiring registration calls for different physical qualifications. For example, performing the duties of a guard may require a higher level of hearing and better eyesight than are necessary for an alarm servicer. Similarly, certain physical deformities or limitations may adversely affect performance as an alarm respondent but have no appreciable effect upon performance as an investigator.

The area of physical qualifications should be carefully studied by the private security regulatory board. These qualifications should become part of their rules and regulations after careful consideration of the relationship between specific duties to be performed by the registrant and any physical problems. Provisions should be made to consider questions of physical competence on an individual applicant basis.

The need for private security personnel to have emotional stability and sound judgment is apparent because of their important roles in maintaining order and protecting lives and property. Whether a person is guarding a remote rock quarry, patrolling a residential area, or investigating business losses, a certain level of mental competence is required. This does not infer that a specific level of educational accomplishment alone would qualify the individual; some people with high school diplomas possess neither commonsense nor emotional stability. Persons whose background investigations indicate they possess sound judgment and emotional stability should be allowed to register as security personnel, regardless of their level of formal education.

Measuring or determining mental competence is not easy, particularly when such determinations must be made for thousands of applicants in the initial stages of registration. Two recommended methods that perhaps can be gradually worked into the registration process are psychological tests and interviews by trained professionals. The present limitations of these methods are recognized, but their validity and usefulness may be increased through continued research. Private security regulatory boards, therefore, should study these methods and keep abreast of research so that the best available means of measuring mental competence can be determined and applied.

The need for morally responsible security personnel cannot be argued, but questions over what, in fact, constitutes being morally responsible are likely. One solution for adding preciseness to the term would be to require that no person who has been convicted of a felony or misdemeanor that reflects upon ability to perform security work should be allowed to register. However, in many cases, the regulatory board may find that an individual has a long list of criminal charges that have never resulted in a conviction but the nature and number of charges may indicate that the person is not morally responsible.

Finally, this standard incorporates the specific training recommendations set forth in Standard 2.5. As was pointed out in Chapter 2, training can significantly improve the competence of security personnel to aid in crime prevention and control. Training is one of the most common areas in the private security industry needing the most improvement. However, unless requirements are mandated by law, the majority of private security personnel may never receive the necessary training. The benefits of training to employers, private security workers, consumers of security services, and the public are too great to be left to the option of employers or individual workers.

It would, of course, be preferable if all security personnel met stringent, professional requirements. However, this report recommends that the initial government-mandated qualifications should be minimum. It is impossible to determine the number of people who would have to register with the private security regulatory boards. It is also impossible to accurately assess the impact of strict qualifications upon the industry. If the requirements are too high and cannot be met by those applying for registration, a serious shortage of available manpower could occur, adversely affecting the industry and those who seek to use it for protection. Thus, the requirements set forth in this standard are minimal but designed as an initial step for eliminating undesirable applicants. The regulatory board should constantly evaluate the requirements. If a particular requirement is too restrictive and is keeping competent and ethical persons out of the field, that requirement should be eliminated. Likewise, if serious problems are occurring that could be corrected by a different or more stringent requirement, it should be added. Although constant evaluation requires maintenance of records and careful analysis, such efforts are necessary in order to balance the interests of the private security industry and society.

Selected References

1. Brennan, Dennis T. *The Other Police.* Cleveland, Ohio: Governmental Research Institute, 1975.

2. Harrigan, James F., Mary Holbrook Sundance, and Mark L. Webb. "Private Police in California: A Legislative Proposal," *Golden Gate Law Review.*

3. Institute for Local Self Government. *Private Security and the Public Interest.* Berkeley, Calif.: Institute for Local Self Government, 1974.

4. Kakalik, James S., and Sorrell Wildhorn. *Private Police in the United States: Findings and Recommendations,* Vol. I, R—869/DOJ. Washington, D.C.: Government Printing Office, 1972.

5. National Advisory Commission on Criminal Justice Standards and

Goals. *Report on Police.* Washington, D.C.: Government Printing Office, 1973.

6. Oglesby, Thomas W. "The Use of Emotional Screening in the Selection of Police Applicants," *Police*, January-February 1958.

7. President's Commission on Law Enforcement and Administration of Justice. *Task Force Report: The Police*, Washington, D.C.: Government Printing Office, 1967.

Standard 11.3
Qualifications for Armed Security Personnel

Every applicant who seeks registration to perform a specific security function in an armed capacity should meet the following minimum qualifications:

1. Be at least 18 years of age;

2. Have a high school diploma or pass an equivalent written examination;

3. Be mentally competent and capable of performing in an armed capacity;

4. Be morally responsible in the judgment of the regulatory board;

5. Have no felony convictions involving the use of a weapon;

6. Have no felony or misdemeanor convictions that reflect the applicant's ability to perform a security function in an armed capacity.

7. Have no physical defects that would hinder job performance; and,

8. Have successfully completed the training requirements for armed personnel set forth in Standards 2.5 and 2.6.

Commentary

Some of the most serious problems in the private security industry are caused by the use of weapons. Throughout this report, various tragic examples have been cited in which injury or death resulted from weapons abuse. Other private security studies have cited similar incidents. Although no statistics are available to determine the frequency of these incidents, it remains unquestioned that the carrying of a firearm includes the potential for serious and dangerous consequences.

Armed security personnel take on an awesome responsibility. Split-second decisions with lethal weapons can result in death or serious injury, and the lives of armed security workers are constantly endangered. Walter M. Strobl stated, in "Private Guards: Arm Them or Not," "the very fact that a weapon is visible will cause the criminal to assume a more violent attitude that could trigger the most violent actions."

Many responsible individuals within the private security industry have long argued against arming security personnel. Proprietary security executives have encouraged executives within their organizations to abandon the use of weapons, and contract organizations have discouraged consumers from requesting armed personnel. One large contract company actually offers incentives to sales personnel who set up contracts that do not require armed personnel; this action should be commended and encouraged.

It is a sad [but] true reflection on our society that some situations require the arming of certain security personnel. It would be foolish in situations in which lives are under constant threat to forbid the use of firearms. But it is not unwise to place firm restrictions on the use of firearms and equally firm requirements on those who are allowed to carry them. For this reason, higher qualifications are established for those who seek registration as armed security personnel than for those who would be unarmed.

A minimum age requirement of 21 years for persons desiring to be registered as armed personnel was first considered. However, it is believed that there is little correlation between maturity, good judgment, and age. An applicant who can meet all of the other requirements should be allowed registration, regardless of age. Therefore, this report recommends that a minimum age requirement of 18 years should be established for registration of armed personnel.

In the area of educational requirements, a higher level should be required for armed registrants than for other security personnel. The basic education qualifications can be met by a high school diploma or by an equivalent written examination designed to measure basic educational aptitudes.

The qualification for mental competence can enable the board to determine if the applicant is able to understand and perform security functions in an armed capacity. A written examination designed to measure the knowledge and skills required or the psychological makeup of the applicant should be used. This qualification is given along with the education requirement, because it is recognized that such formal education or equivalent does not automatically indicate a person is psychologically capable of carrying a weapon.

Although almost totally ignored by both existing State regulatory boards and by private security employers, psychological testing to screen out the obvious cases of emotionally unstable or unsuitable persons should be an important and integral portion of the competency requirement. This step could prevent psychopaths or other seriously mentally ill persons from being certified as armed guards.

It is difficult to list the specific acts that would indicate that an individual was not morally responsible to carry a weapon. Sometimes a person may meet the listed qualifications, but a review of his records may indicate a very questionable background. A long list of criminal charges or a series of jobs that ended in firing would perhaps be incidents to watch for. Because of the requirement for a hearing before denial, this requirement is not believed to be too general. Any applicant denied registration on this ground would have an opportunity to be heard and to show if the decision was arbitrary and capricious.

Any person who has been convicted of a felony involving the use of a weapon should not be registered in an armed capacity. No exception should be made, regardless of extenuating circumstances, passage of time, or indications of rehabilitation. The responsibility of carrying a firearm is too grave to take the chance that a person previously misusing a gun would not do so again.

Although no flexibility is recommended for felony convictions involving weapons, convictions for other offenses should be carefully studied before denying registration. It is in the public interest to assist the rehabilitation of convicted offenders by removing restrictions upon their ability to obtain employment. But it also must be recognized that the ex-offender is being registered to perform a security function in an armed capacity. Therefore, if an applicant has a conviction record, the regulatory board should carefully consider whether such convictions reflect upon the applicant's ability to perform a security function in an armed capacity. In making its determination, the regulatory board should consider the following:

1. The specific security function the applicant is registering to perform;
2. The nature and seriousness of the crime;
3. The date of the crime;
4. The age of the applicant when the crime was committed;
5. Whether the crime was an isolated or repeated incident;
6. The social conditions that may have contributed to the crime; and
7. Any evidence of rehabilitation, including good conduct in prison or in the community, counseling or psychiatric treatment received, acquisition of additional academic or vocational schooling, successful participation in correctional work-release programs, or the recommendation of persons who have, or have had, the applicant under their supervision.

The next qualification concerns physical requirements. Such requirements should not be unduly restrictive and should not include height and weight specifications or other requirements that have little relation to performance in an armed capacity. Physical standards, however, cannot be totally disregarded. Obviously, good eyesight and hearing are vital to anyone who carries a weapon. In order to protect the individual and the public, specific vision and hearing requirements should be carefully considered and delineated by the regulatory board.

Finally, this standard incorporates the specific training recommendations set forth in Standards 2.5 and 2.6. As pointed out in Chapter 2, training can greatly improve the competence of security personnel to aid in crime prevention and control but needs perhaps the most improvement of any private security area. However, unless requirements are mandated by law, the majority of private security personnel may never receive such training. The benefits gained through training to employers, private security workers, consumers, and the public are too great to be left to the option of employers or individual workers. Competence and proficiency in the use of a firearm should be demonstrated by those who seek to be registered as armed security personnel. This can best be shown by successful completion of the required arms training recommended in Standards 2.5 and 2.6.

In summary, any individual allowed to carry a weapon needs to be able to make decisions that require mature, calculated, and sound judgment. The armed security worker should also possess the physical and emotional makeup to act with split-second timing, if necessary, and be thoroughly trained in the use and legal implications of the weapon to be carried. Considering the life-or-death potential involved, every effort must be made to prevent any but the most qualified and capable individuals from performing in an armed capacity. Although there is no magic panacea to ensure that a life will not be taken accidentally or unnecessarily, the risks involved demand strict qualifications for registration of armed personnel.

Selected References

1. Brennan, Dennis T. *The Other Police*, Cleveland, Ohio: Governmental Research Institute, 1975.

2. Harrigan, James F., Mary Holbrook Sundance, and Mark L. Webb. "Private Police in California: A Legislative Proposal," *Golden Gate Law Review*.

3. Institute for Local Self Government. *Private Security and the Public Interest*. Berkeley, Calif.: Institute for Local Self Government, 1974.

4. Kakalik, James S., and Sorrel Wildhorn. *Private Police in the United States: Findings and Recommendations*, Vol. 1, R-869/DOJ. Washington, D.C.: Government Printing Office, 1972.

5. New Jersey State, P.S. 1968, c. 282 (C.2A: 168A).

6. Strobl, Walter M. "Private Guards: Arm Them or Not?" *Security Management*, January, 1973.

Appendix B
A Commentary and Checklists for Security Surveys

Security Surveys Checklists

At the start, physical security surveys were developed for defense industrial plants and military installations. It was early recognized by civilian police personnel that the survey technique could be applied to areas, buildings, or activities of all types to identify the conditions of property or the interaction of people most conducive to criminal activity. Some police and security personnel have tried to maintain a distinction between the physical security survey and the crime prevention survey. In actuality, both types of surveys have the same goals: the identification of risks and the recommendation that those risks be eliminated or minimized through the employment of security measures. As noted in Chapters 4 through 9, physical barriers, lights, alarms, guards and security awareness programs for all employees can materially reduce the risk to lives and loss of property or proprietary information.

Initially, security surveys were conducted by experienced criminal and intelligence investigative personnel. Their job was to inspect various facilities and recommend installation of some of the basic protective devices. Shortly the precursor of the Defense Department established some minimal standards for fences, lighting, alarms and guards. Because security measures had to meet those standards, long lists of the requirements were prepared in advance of the inspections. Thereby the inspectors' jobs were made easier, despite the need to prepare a narrative report of the inspection. This involved a lot of writing, and few inspectors enjoyed the tedium. In time, checklists evolved from those long lists of standards and some security survey reports were accepted with little narrative information added.

Over time, the basic security survey checklist was found to have some drawbacks. It did provide a general plan to inspect a facility of several buildings or a separate activity. It also provided lists of general questions concerning fences, lighting and guards, as well as other aspects of security. But as unique facilities with unusual security problems required surveys, the original checklists were found lacking. Specialized checklists had to be devised to meet the new requirements. Hence, the development of checklists for retail, bank and EDP operations, to name a few.

It seems that every checklist contains the warning: "This checklist is merely a guide and may not cover all the areas needed for every specific survey." It did not take long to decide that a checklist devised for one specific building, activity or facility might not adequately address the security problems in the next building, activity or facility. Only the chain type fast food or retail stores that are built in one style and require strict adherence to standard operational policies might be suitably inspected by a standard checklist. Still, all checklists have room for additional remarks when warranted.

Qualifications of the Surveyor

Modern day security survey checklists have not eliminated the need for a well-trained and experienced surveyor. During the early days of security surveying, the criminal or intelligence investigator generally had extensive field experience, as well as the benefit of military service schools that specialized in security. With the establishment of the National Crime Prevention Institute at the University of Louisville in the early 1970s, civilian-oriented crime prevention courses and instruction in security surveys were offered to enrolled police students. Since then, security surveys have been taught as a semester-long course in colleges or by three- to five-day-long courses offered by private organizations such as the American Management Association and the American Society for Industrial Security. Recent references to the qualifications of surveyors do not reflect any change over the years. Surveyors are still recommended to be well-trained and experienced.

A well-trained and experienced surveyor cannot be calculated by the number of training courses attended, the college credits accumulated, or by the number of years of employment by a police or security department. A well-trained and experienced surveyor is the person who takes to the job site the knowledge and understanding that crimes are committed by all kinds of people, young and old, male and female. These people come from all walks of life: congressmen, judges, police officers, ministers, mechanics and laborers. Some people have illegally entered facilities or premises using helicopters, trucks, ladders, overhanging tree limbs and false identification. Crimes ranging from murder, theft, industrial espionage and vandalism have been committed inside those premises, and, when caught by the authorities, those criminals, more often than not, denied the acts and threatened a legal suit for any violation of their legal and civil rights. A well-trained and experienced surveyor needs to know these things and how to cope with them satisfactorily. He should not limit himself by thinking like a criminal to prevent criminal activity; he should think better than that criminal to prevent the crime.

Texts such as this one and survey checklists reflect the research and experience of many surveyors and authors. These are excellent starting points for a surveyor. Yet he must be ever mindful of the caveat earlier explained: "This checklist is merely a guide and may not cover all the areas needed for every specific survey." An open and inquisitive mind will help the surveyor fill in the gaps left in this text and in standard checklists. When the surveyor can perform these tasks with confidence, then he is well-trained and experienced.

Conducting the Survey

Chapter 10 clearly explains the administrative procedures for a security survey. Upon completing the tasks outlined in the section titled *Preliminary Activities*, a surveyor should have a grasp of the size and location of the facility, the number of persons who may be encountered, and the scope of the daily operations. From this preliminary study tentative checklists can be assembled. Yet, it must be remembered that many more items may have to be added to that tentative checklist.

After an orientation tour of the facility, and all subordinate supervisors know of the intent of the survey, the real work begins. A basic or general type survey checklist, such as the one included in this appendix, can be used while the surveyor visits and observes every physical location of the facility. Preferably, the survey is started from the entry area outside, until every part of the facility has been inspected. The checklist can be expanded when needed.

Most surveyors will develop an ability to estimate sizes, heights and numbers of objects with some accuracy and use these for their reports. That is not a good idea. A survey report needs accuracy in every respect. Therefore, a surveyor should be familiar with and carry tape measures, chalk, paper for sketches and a camera.

Some surveyors walk through buildings and point and ask, "Is that door locked?" Given an affirmative answer, the door is not checked. Only if it is checked can it be known for certain that it was in fact locked. So never take for granted what appears to be obvious. And never accept for a fact any description or explanation by an employee or supervisor. They will often intentionally lie or mislead the surveyor, because they want their section to look good on the report, or because they are covering up some problem.

The first visit of the facility should be during daylight hours to better see the physical plant and learn what is normal. Some facilities burn outdoor lights twenty-four hours a day, otherwise a night visit is necessary to see if all the lights work. Some gates are opened only for shift changes. Sometimes rail access gates are never closed or are guarded only at night and are used by shortcutting employees. Such valuable information can be learned during the daytime.

As mentioned earlier, all doors, gates, bins or other objects that are purportedly locked should be physically checked. If there is a padlock on the hasp, is it locked? If the padlock is hanging on an opened hasp, is the padlock locked so that it cannot be replaced temporarily with a look-a-like by a thief? If the window is reported as always locked, are there signs that it had been opened recently? Are the bars over the windows secure, or are they so rusty they would fall off if the surveyor pulled on them? If an exterior light bulb is loosened, how long will it be inoperative? Questions that come to the surveyor's mind during the survey should at least double the items on any checklist.

The night time survey of a facility encompasses the same physical areas in an effort to determine whether there are security weaknesses that occur only at night or are not obvious during the day. If the facility is non-operational at night, are there guards or janitorial personnel on the premises? Do off-duty employees or vendors return at night? Is equipment or property stolen or used by these visitors? Do guard personnel entertain friends while on duty? Are all security policies enforced? Again, any imaginative surveyor could double the checklist during a night time inspection.

Coordinating Interviews

In order to engender support for this and future surveys and the overall security program, the discrepancies or suggestions should be reviewed with the appropriately responsible supervisor before the formal report is submitted. There must be an emphasis that the survey is not a punitive inspection, rather, it is a joint effort to prevent losses and promote profitability. Often the lower level supervisors can be given credit for their input and cooperation; this may promote further individual support so necessary for the security awareness efforts after the surveyor is gone from the facility.

The Follow-up Survey

Another important reason to maintain a businesslike yet friendly relationship with the facility personnel is that those people will have to take the corrective actions if the discrepancies are to be remedied. If they are not remedied, the surveyor's efforts are wasted. One method to determine if the deficiencies are corrected is to perform a follow-up survey. This survey is normally undertaken 30, 60 or 90 days after the initial survey is reported. Only those areas with reported deficiencies are checked, but sometimes new concerns predicate the expansion of the follow-up survey to include newly developed problem areas. The report of the follow-up survey generally includes only the list of discrepancies and a description of the remedial action undertaken. There have been occasions where several follow-up surveys were requested by management before corrective steps were taken to rectify the discrepancies. This again demonstrates the need to develop a cooperative relationship with the supervisors affected by the survey. Without their support, no security program will succeed.

Frequency of Survey

As stated in Chapter 10, there is no magic number or time frame to conduct security surveys. New or changed facilities or activities dictate a survey. So would continual losses, damages or injuries. But often security surveys will not solve the existent underlying problems that cause the losses, damages or injuries. At such times the surveyor must suggest that a security or safety awareness program would be more effective to stem the increase of unwanted incidents. During normal surveys there will be a lot of personal contact with supervisors and employees. It does not take long to distinguish between a healthy competitive spirit and a lingering animosity. People in both groups talk about what they have on their minds. Low pay, swing shifts, partiality of supervisors and poorly explained rules frequently are topics mentioned by employees caught stealing or damaging merchandise or equipment. There is no final number of surveys that will solve those problems, but a management attuned to the problems can reduce the inherent problems by an educational program. Any such program that will reduce losses, damages or injuries is truly an effective security program.

Security Survey Checklists

Included in this appendix are two sample security checklists. Most of the items in the checklists have been handed down from surveyor to author to surveyor. Use these items freely, but do not limit a survey to these few specific items. Even a surveyor with limited training and experience should be able to double each list while on the job.

BASIC SECURITY SURVEY

Preparing agency:_____ Name and address of facility surveyed:_____

Date of survey:_____ Date of report:_____ Date of previous report:_____

Name and title of person responsible for facility:_____

Name and title of Security Supervisor:_____

No. of security personnel:_____ No. of employees:_____

No. of acres:_____ Total sq. ft. of building space:_____

No. of bldgs:_____ No. of rooms:_____ No. of floors in tallest bldg._____

No. of miles of roadway:_____ No. of visitor parking spaces:_____

No. of registered vehicles:_____ No. of vendor parking spaces:_____

Total no. of parking spaces:_____

A brief history and description of the business carried on at the facility:

Name of person making survey (typed):_____
Signature of person making survey:_____
Name of person making report (typed):_____
Signature of person making report:_____

Part I — Facility Environment

	Yes	No	Remarks
1. Do employees feel secure at this location?			
2. What is the crime rate?			
3. Can local police observe approaches to the facility?			
4. Do other buildings & structures present security hazards?			
5. Does landscaping or shrubbery present a security hazard?			

	Yes	No	Remarks
6. Do trees, poles or fences offer easy access to the roof?			

Part II — Perimeter

	Yes	No	Remarks
7. Is facility surrounded by a fence or other barrier?			
8. Are fences properly constructed with an outrigger top guard?			
9. Height of fence?			
10. Describe fence construction.			
11. Is selvage twisted at top and bottom of fence?			
12. Is bottom of fence within two inches of solid ground?			
13. Is the fence other than chain link?			
14. If perimeter barrier is constructed of stone or other masonry, what is its height?			
15. If a wall, is it protected at the top by a proper guard of wire or broken glass?			
16. Are perimeter barriers increased in height at junctions with buildings and other critical points?			
17. Are barriers inspected for defects? If so, by whom and how often?			
18. Are openings (culverts, manholes) which are 96 sq. in. or larger protected by mesh or wire?			
19. Is mesh no greater than 2 sq. in.?			
20. How many gates and entrances are in the perimeter?			
21. Are all perimeter entrances secured with locking devices?			
22. Are all entrances closed and locked when not in use?			
23. Are perimeter openings inspected by guards for security?			
24. Are warning signs posted at all entrances?			

	Yes	No	Remarks
25. Are "No Trespassing" signs posted to be observed for at least 50 yards?			
26. Are clear zones maintained on both sides of perimeter area?			
27. Is parking allowed against or close to perimeter barrier?			
28. Do guards patrol perimeter area?			
29. Are perimeter barriers protected by intrusion alarm devices?			

Part III — Exterior Lighting

30. Does facility use municipal lighting? Is it dependable?
31. What type of lighting is used?
32. Are night lights activated automatically?
33. What is the plan for replacing burned out lights?
34. Is there adequate lighting around buildings, company vehicles, and cargo?
35. Are customer and employee parking lots lighted sufficiently?
36. Does the lighting provide adequate illumination over perimeter and entrances?
37. Is there an auxiliary source of power for lighting?
38. What is plan for standby or emergency lighting?
39. Does the emergency lighting activate automatically when needed?

Part IV — Doors

40. Do all doors lock from both sides except for the main entrance door?
41. Are all unlocked doors properly protected?
42. Are the doors, locks and hardware in good repair?
43. Are the exterior doors strong?

Yes No Remarks

44. What type of lock is used on doors?
45. Are electrically-operated overhead doors locked when not in use?
46. Are overhead doors operated by rollers on tracks sufficiently strong?
47. Are unnecessary doors bricked or permanently sealed?

Part V — Windows

48. What type of glass is used in the windows?
49. How are windows located less than 18 ft. from the ground protected?
50. Are the windows more than 14 ft. from trees, poles, etc.?
51. Is valuable merchandise visible through the windows?
52. If windows are connected with an alarm system, what type?
53. Are unnecessary windows bricked and sealed shut permanently?

Part VI — Keying System

54. Is there a key control officer?
55. Are all locks and keys supervised and controlled by the key control officer?
56. Are personnel required to produce their keys periodically?
57. Is a dependable person responsible for the master keys?
58. Are key holders allowed to duplicate keys?
59. Are keys marked, "Do not duplicate"?
60. Are keys issued to anyone other than installation personnel?
61. Is the removal of keys from the premises prohibited?
62. Are files kept on the buildings and entrances for which keys are issued?
63. Are files kept on the number and iden-

	Yes	No	Remarks
tification of keys issued?			
64. Are files kept on the location and number of master keys?			
65. Are files kept on the location and number of duplicate keys?			
66. Are files kept on the location and number of keys held in reserve?			
67. Are files and keys kept in a locked, fireproof container?			
68. Is the fireproof container kept in an area of high security?			
69. Are losses and thefts of keys promptly investigated by the key control officer?			
70. Must requests for reproduction or duplication of keys be approved by the key control officer?			
71. Are locks changed when keys are lost or stolen?			
72. Are locks rotated within the facility at least annually?			

Part VII — Storage Areas Outside Building

73. Are dangerous materials or chemicals stored outside the building area?
74. Is the area protected by a fence?
75. Are trespassing signs posted?
76. Are the areas adequate for the materials being stored?
77. Are these areas locked and secured?
78. Are these areas illuminated?
79. Are these areas patrolled?

Part VIII — Employee Lockers

80. Are lockers provided to all employees?
81. What type of lock is used on lockers?
82. Does the company have a key to enter employee lockers?
83. Does the company have written consent

	Yes	No	Remarks
from the employee to open his locker at any time?			
84. Are the lockers located away from removable merchandise?			
85. Are regular and unscheduled inspections made of the lockers?			
86. What is the company policy if stolen merchandise is found in a locker?			
87. Do employees have access to each others' lockers?			

Part IX — Outside Parking for Employees or Customers

88. Is parking in a designated area outside the perimeter?
89. Are parking areas patrolled?
90. Are parking areas well lighted?
91. Is parking allowed near a loading dock?
92. Is there any barrier between loading docks and parking area?
93. Where are company vehicles parked in keeping with good security practices?

Part X — Fire Protection

94. Has a fire department ever surveyed this facility?
95. Were there any recommendations?
96. Were these recommendations followed?
97. Does the facility comply with fire regulations and ordinances?
98. Are fire doors protected with panic bars and door alarms for emergency use?
99. Does the facility have a fire safety program?
100. Have employees been adequately trained and drilled on the fire procedures?
101. Do employees know where fire equipment is located?
102. Are fire extinguisher locations distinctly marked?

Yes No Remarks

103. Is the fire department number posted by all telephones?
104. Is there a sprinkler system?
105. Is the sprinkler system inspected regularly?
106. What is the average response time for the fire department to arrive at the facility?
107. Is there adequate water pressure at the facility?
108. Are signs posted for procedures to follow in case of fire?
109. Do employees know what to do in case of fire?

Part XI — Guard Force

110. Is there a security or guard force?
111. Is it adequate for the security and protection needed?
112. Is the force reviewed periodically to ascertain its effectiveness?
113. Can the guards use the communications properly?
114. Do the guards meet minimum qualification standards?
115. Are the guards on duty armed? Describe weapons.
116. Are the weapons inspected periodically?
117. Are guards required to complete basic courses in firearms?
118. Is there in-service training on security and firearms?
119. Does each guard carry a flashlight?
120. Do the activities of the guards follow established policy?
121. Does each guard write a daily report?

Part XII — Personnel Identification and Control

122. Is an identification card or badge used?

Yes No Remarks

123. Is a picture of the employee on the badge?
124. Is the picture updated?
125. Is there a Standard Operating Procedure for the identification system?
126. Are personnel knowledgeable about the system?
127. Is special identification required for high-security areas?
128. Are visitors issued a visitor's pass or badge?
129. Can this badge be used only in a designated area?
130. Is everyone required to wear an ID badge at all times in the facility?
131. Do guards at entrances and exits compare the badge photo to the bearer?
132. Are badges recorded and controlled by an accountability procedure?
133. Are replacement badges identifiable as such?
134. Are temporary badges used?
135. Are rosters of lost badges posted at guard control points?
136. Are badges distinct in appearance for different areas?
137. Do procedures insure the return of ID badges upon termination of employment or transfer?
138. Is ID system under the supervision and control of security officers?
139. Is there a visitor escort policy?
140. Are visitor arrivals recorded?
141. Must a visitor's ID badge be displayed at all times?
142. Are visitors allowed to move about the facility unattended?
143. Must visitors turn in ID passes when leaving the facility?

Yes No Remarks

144. Is visitor departure time recorded?
145. What is the procedure if a visitor fails to turn in an ID pass?
146. Are permanent records of visitors maintained?
147. Are restriction notices displayed prominently at appropriate entrances?
148. Is there inspection of all packages and materials carried in and out of the facility?

Part XIII — Sensors and Switches

149. Are there sensors and switch devices?
150. Where are they located?
151. Are the security devices adequate for the degree of security required?
152. Is there an operation of a photographic or CCTV identification camera in security-sensitive areas?
153. Are security devices adequately protected against attacks?
154. Are these security devices inspected regularly to insure working condition?
155. Are the security devices connected to a silent alarm?
156. Are local alarms loud enough to alert a civic-minded person in the area?
157. Would a police force nearby respond to a sounding alarm?
158. Who is responsible for the security devices at the facility?
159. Is there an auxiliary power source for the alarm system?
160. Are all wires for the alarm system underground so they are tamper-resistant?
161. Are the alarm systems designed, and are locations recorded, so repairs can be made rapidly in an emergency?

	Yes	No	Remarks
162. Is there someone available at all times to make repairs to the alarm system?			
163. Is there someone available at all times to make repairs to the communications system?			
164. Note other deficiencies or irregularities:			
165. Note security measures implemented but not listed above:			
166. Recommendations:			

SECURITY SURVEY OF MOTEL/HOTEL

This specialized checklist is designed to supplement the Basic Security Survey Checklist.

Name and address of facility: ______________________________

Facility owned by: ______________________________

Facility official coordinating survey: ______________________________

Date of Survey: _______ Date of report: _______ Previous survey: _______

Name and organization of surveyor: ______________________________

	Yes	No	Remarks
1. Are entrances to non-public areas marked "Employees only"?			
2. Are I.D. badges issued to employees?			
3. Are entrances from outside monitored and access controlled?			
4. Is a package check system used?			
5. Are salesmen and vendors screened by management before being allowed entrance?			
6. Are goods delivered to the loading platform and/or premises properly received?			
7. Do supervisors inspect trash removal procedures?			

	Yes	No	Remarks
8. Do all emergency exits provide a clear path of egress?			
9. Is a system used to control stock of alcoholic beverages?			
10. Are bars stocked with a predetermined quantity of each item?			
11. Is the quantity of each item at the bar regularly counted?			
12. Are bar receipts and cash checked for accuracy and honesty?			
13. Are alcoholic beverages stored separately from other supplies?			
14. Is the main alcoholic beverage storeroom a high security area?			
15. Are bars always attended or secured by hardware?			
16. Are hotel liquor bottles marked in some way to prevent "Bar Padding" by bar tenders?			
17. Is there dual control for taking liquor inventory?			
18. Are sales slips or records provided for each drink served?			
19. Are sponsors of parties warned to keep account of liquor consumption to prevent "padding" by dishonest employees?			
20. Are bar employees trained in applicable laws governing alcoholic beverages?			
21. Are patrons who are obviously drunk denied additional liquor sales?			
22. Are age limit regulations concerning the sale of alcohol strictly adhered to? Are appropriate signs posted?			
23. Have meat and produce purchasing specifications been established?			
24. Are incoming meat and produce orders inspected for weight and freshness?			
25. Are goods from storerooms and kitchen dispersed only upon signature by an authorized person?			

	Yes	No	Remarks
26. Do kitchen employees eat food on the spot?			
27. Do supervisors inspect areas where food could be hidden for a planned theft?			
28. Is a cashier responsible for the pricing and totaling of checks and handling all money?			
29. Is surplus food incorporated in future dishes to cut down on waste in preparation?			
30. Is garbage controlled to prevent theft?			
31. Are all laws affecting restaurant operations strictly adhered to?			
32. Are claims of illness by customers thoroughly investigated?			
33. Are credit cards checked against "hot sheets"?			
34. Are measures used to protect the cash flow of the hotel?			
35. Is the hotel's safe combination changed regularly and access restricted?			
36. Are cash and change funds adequately secured and are they checked at unannounced intervals?			
37. Are vending machines cleared of their cash on a routine basis and checked by the head cashier?			
38. Is the money transported to the bank in a secure container, and are different routes and time schedules applied?			
39. Are preventive measures taken to deal with robberies?			
40. Are employees instructed how to act during an armed robbery?			
41. Does the hotel check suspected stolen credit cards?			
42. Is there a policy for cashing checks?			
43. Are cashiers of the hotel checked randomly for accuracy and honesty in cash transactions?			

	Yes	No	Remarks
44. Are cashiers bonded?			
45. Are cashiers trained to check for counterfeit bills, forged checks and bogus credit cards?			
46. Does the hotel provide a theft-resistant, fireproof vault or some type of cabinet for protection of guests' "valuable items"?			
47. Do customers check personal belongings with management?			
48. Is the hotel staff trained in the preventive measures of fraud against the hotel?			
49. Are pass keys issued to maids checked in and out daily?			
50. Are maids told to be alert for possible thieves?			
51. Do maids leave the doors to rooms open when they are being cleaned?			
52. Is an inspection made of a room as soon as it is vacated?			
53. Are secure areas provided for convention displays shipped ahead of time?			
54. Are rooms inspected for personal belongings after meetings and banquets?			
55. Are there certain procedures used for guests who are thought to be "skippers"?			
56. Do maids keep utility and storage closets locked?			
57. Is linen stored in a secure location?			
58. Is linen marked with adequate identification?			
59. Is it possible to determine what part of the linen cycle is losing the greatest amount of goods?			
60. Are parking garages patrolled?			
61. Is the parking lot enclosed by a fence to discourage trespassers?			
62. Are damages to guests' cars carefully investigated?			

	Yes	No	Remarks
63. Is the hotel responsible for the contents of guests' cars?			
64. Is recovered, lost, abandoned, and unclaimed property accounted for and properly secured according to policy?			
65. Is there a written fire prevention policy?			
66. Are all fire exits marked clearly?			
67. Do all employees know what to do in case of fire?			
68. Is there a disaster plan?			
69. Have the employees been trained in how to handle bomb threats?			
70. Does the hotel security have a V.I.P. protection plan?			
71. Is there a written plan for strikes, civil unrest, or any other type of disturbances?			
72. Do the employees and security officers know how to detect drug users, loan sharking, illegal gambling, and prostitution on the premises?			
73. Are management employees instructed in how to deal with criminal activity?			
74. Are all laws pertaining to the business clearly understood by the employees and are references readily available?			

Appendix C
College Security Program List*

(All programs verified in 1985)

Reprinted from Volume 9 (1986), Number 1, *Journal of Security Administration.* Norman R. Bottom, Jr., Ph.D., Editor; London House Press, 1550 Northwest Highway, Park Ridge, IL 60068; (312) 298-7311

Code for College Degrees

AA	Associate of Arts
AAB	Associate of Applied Business
AAS	Associate of Applied Science
AS	Associate of Science
BA	Bachelor of Arts
BCA	Bachelor of Career Arts
BPS	Bachelor of Public Safety
BS	Bachelor of Science
CERT	Certificate
(Con)	Concentration
DIP	Diploma
Ed.S.	Educational Specialist
(Emp)	Emphasis
LE	Law Enforcement
LSA	Law & Security Administration
MA	Master of Arts
MCJA	Master of Criminal Justice Administration
MPM	Master of Professional Management
MS	Master of Science
(Opt)	Option
(Spec)	Specialist

ALABAMA

Auburn University of Montgomery	BS(Opt), MS(Opt)
Chattahoochee Valley College, Phenix City	AAS

ARIZONA

Navajo Community College, Tsaile	CERT
Northern Arizona University, Flagstaff	BS(Opt), MS(Opt)

CALIFORNIA

California State University, Long Beach	CERT, BS, MS(Minor)
Cerritos College, Norwalk	CERT, AA(Spec)
De Anza College, Cupertino	AA(Opt), CERT
East Los Angeles College, Monterey Park	AA, CERT
Golden Gate University, San Francisco	BS, CERT
Golden West College, Huntington Beach	AA, CERT
Monterey Peninsula College, Monterey	AS(Con)
Mount San Antonio College, Walnut	AS(Opt)
Palomar Community College, San Marcos	AA(Spec), CERT
Saddleback College, Irvine	CERT

COLORADO

Red Community College, Golden	CERT

CONNECTICUT

Housatonic Community College, Bridegport	AS(Opt)
Tunxis Community College, Farmington	CERT
University of New Haven, West Haven	CERT, BS

DISTRICT OF COLUMBIA

George Washington University	MA

FLORIDA

Barry University, Miami Shores	MBA
Daytona Beach Community College, Daytona Beach	AS, AA
Manatee Community College, Bradenton	AS

ILLINOIS

Belleville Area College, Belleville	AAS
Lewis and Clark Community College, Godfrey	CERT, AAS
Lincoln Land Community College, Springfield	CERT, AAS
Loop College, Chicago	CERT, AS
Moraine Community College, Palos Hills	AAS
Thornton Community College, South Holland	AAS, CERT
Western Illinois University, Macomb	BS(Minor)
William Rainey Harper College, Palatine	CERT

INDIANA

Indiana State University, Terre Haute	AA(Emp), BA(Emp), MA(Emp), MS(Emp)

Institution	Degree
Indianapolis Vocational Technical College, Ft. Wayne	CERT
University of Evansville, Evansville	CERT

IOWA

Institution	Degree
Southeastern Community College, West Burlington	AA(Emp)
St. Ambrose College, Davenport	BS(Con)

KANSAS

Institution	Degree
Wichita State University, Wichita	BS(Spec)

KENTUCKY

Institution	Degree
Eastern Kentucky University, Richmond	AA, BS, MS(Opt)

MARYLAND

Institution	Degree
Catonsville Community College, Catonsville (Baltimore County)	CERT, AA
Community College of Baltimore, Baltimore	CERT, AA
Coppin State University, Baltimore	BS(Minor)
Montgomery College, Rockville	AA(Opt)

MASSACHUSETTS

Institution	Degree
Northeastern University (College of Criminal Justice), Boston	BS, MS
Northeastern University (University College — L. E. Programs), Boston	AS, BA

MICHIGAN

Institution	Degree
Ferris State College, Big Rapids	BS(Opt)
Jackson Community College, Jackson	CERT, AAS, AS(Con)
Lake Superior State College, Sault Ste. Marie	BS(Emp)
Macomb County Community College, Mt. Clemens	CERT, AAS
Madonna College, Livonia	CERT, AA, BA
Michigan State University, East Lansing	BS(Con), MS(Con)
Northern Michigan University, Marquette	BS
Oakland Community College, Auburn Hills	AAS(Retail)
Schoolcraft College, Livonia	AAS
University of Detroit, Detroit	MS

MISSOURI

Institution	Degree
Central Missouri State, Warrensburg	BS, MS, Ed.S.(Emp)
Columbia College, Columbia	BA(Con), BS(Con)
Missouri Southern State College, Joplin	BS(Con)
Tarkio College, Tarkio	CERT, BS

NEBRASKA

Metropolitan Technical Community College, Omaha	AS

NEVADA

Clark County Community College, Las Vegas	AAS(Spec)

NEW JERSEY

Essex County College, Newark	CERT
Jersey City State College, Jersey City	BS
Union College, Cranford	AA(Opt)

NEW YORK

Hudson Valley Community College, Troy	AAS
Iona College, New Rochelle	CERT
John Jay College of Criminal Justice, New York	AS
Long Island University, Greenvale	BA, MPS
Mercy College, Dobbs Ferry	CERT, AS(Emp), BS
Monroe Community College, Rochester	CERT, AAS (Computer Security)
Nassau Community College, Garden City	AS
Orange County Community College, Middletown	CERT
Private Industry Loss Control School, New York	CERT
St. John's University, Jamaica	BS
Westchester Community College, Valhalla	CERT, AAS

NORTH CAROLINA

Appalachian State University, Boone	BS(Con)
Central Piedmont Community College, Charlotte	AAS, CERT
Mayland Technical Institute, Spruce Pine	AAS(Opt)
Nash Technical College, Rocky Mount	AAS(Emp)
Surry Community College, Dobson	AAS (Protective Services Tech)

NORTH DAKOTA

Bismarck Junior College, Bismarck	AA

OHIO

Case Western Reserve University School of Law, Cleveland	CERT
Choffin Career Center Vocational High School, Youngstown	CERT
Cincinnati Technical College, Cincinnati	CERT, AAB
Cuyahoga Community College, Cleveland	AA
Hocking Technical College, Nelsonville	AAS
Jefferson Technical College, Steubenville	AAS(Emp)

Lorain Community College, Elyria	AAS
Ohio University, Chillicothe	AAS
Ohio University (Lifelong Learning/ Independent Study), Athens	AAS
Owens Technical College, Toledo	AAS
Sinclair Community College, Dayton	AAS
Tiffin University, Tiffin	CERT

OKLAHOMA

Oklahoma City University, Oklahoma City	MCJA(Emp)

OREGON

Clackamas Community College, Oregon City	CERT
Lane Community College, Eugene	AS(Emp)
Portland Community College, Portland	CERT

PENNSYLVANIA

Alvernia College, Reading	BA(Minor)
Community College of Allegheny County, Monroeville	AS
Luzerne County Community College, Nanticoke	AS, AA, CERT
Mercyhurst College, Erie	BA, BA(Minor)
Pennsylvania State University, Fayette Campus, Uniontown	BA(Spec)
University of Pittsburgh, Pittsburgh	BA(Spec), MA(Tract)
Villanova University, Villanova	CERT
York College of Pennsylvania, York	BS(Con), BS(Minor)

SOUTH CAROLINA

Greenville Technical College, Greenville	(Emp)

TENNESSEE

Cleveland State Community College, Cleveland	AS(Opt)
Shelby State Community College, Memphis	CERT
Walters State Community College, Morristown	AS(Con)

TEXAS

Dallas Baptist University, Dallas	BCA
Houston Community College, Houston	CERT
University of Texas, Arlington	BS(Emp)
University of Texas, San Antonio	BA(Emp)

UTAH

Weber State College, Ogden	CERT, AS

VIRGINIA

Northern Virginia Community College,

Alexandria	AAS
Northern Virginia Community College, Annandale	AAS
Northern Virginia Community College, Manassas	AAS, CERT
Northern Virginia Community College, Woodbridge	AAS, CERT
Virginia Commonwealth University, Richmond	BPS(Opt)

WEST VIRGINIA

West Virginia Northern Community College, Wheeling	AAS

WISCONSIN

Fox Valley Technical Institute, Appleton	AA

AUSTRALIA

Philip Institute of Technology, Cobrug Campus, Melbourne, Victoria	DIP

CANADA

Algonquin College of Applied Arts and Technology, Ottawa, Ontario	CERT, LSA
College of Trades and Technology, St. Johns, Newfoundland	CERT
Concordia University, Montreal, Quebec	CERT
Fanshawe College of Applied Arts and Technology, London, Ontario	CERT
Lethbridge Community College, Lethbridge, Alberta	CERT
Mount Royal College, Calgary, Alberta	CERT, LSA
Sheridan College, Brampton, Ontario	LSA
Sir Sanford Fleming College, Peterborough, Ontario	DIP(Opt)
University of Alberta, Calgary, Alberta	CERT

Index